Advances in Intelligent Systems and Computing

Volume 1444

The series "Advances in Intelligent Systems and Computing" contains publications on theory, applications, and design methods of Intelligent Systems and Intelligent Computing. Virtually all disciplines such as engineering, natural sciences, computer and information science, ICT, economics, business, e-commerce, environment, healthcare, life science are covered. The list of topics spans all the areas of modern intelligent systems and computing such as: computational intelligence, soft computing including neural networks, fuzzy systems, evolutionary computing and the fusion of these paradigms, social intelligence, ambient intelligence, computational neuroscience, artificial life, virtual worlds and society, cognitive science and systems, Perception and Vision, DNA and immune based systems, self-organizing and adaptive systems, e-Learning and teaching, human-centered and human-centric computing, recommender systems, intelligent control, robotics and mechatronics including human-machine teaming, knowledge-based paradigms, learning paradigms, machine ethics, intelligent data analysis, knowledge management, intelligent agents, intelligent decision making and support, intelligent network security, trust management, interactive entertainment, Web intelligence and multimedia.

The publications within "Advances in Intelligent Systems and Computing" are primarily proceedings of important conferences, symposia and congresses. They cover significant recent developments in the field, both of a foundational and applicable character. An important characteristic feature of the series is the short publication time and world-wide distribution. This permits a rapid and broad dissemination of research results.

Indexed by DBLP, INSPEC, WTI Frankfurt eG, zbMATH, Japanese Science and Technology Agency (JST).

All books published in the series are submitted for consideration in Web of Science.

For proposals from Asia please contact Aninda Bose (aninda.bose@springer.com).

Ambuja Salgaonkar · Makarand Velankar
Editors

Computer Assisted Music and Dramatics

Possibilities and Challenges

Springer

Editors
Ambuja Salgaonkar
Department of Computer Science
University of Mumbai
Mumbai, India

Makarand Velankar
Information Technology
MKSSS's Cummins College of Engineering
Pune, India

ISSN 2194-5357 ISSN 2194-5365 (electronic)
Advances in Intelligent Systems and Computing
ISBN 978-981-99-0886-8 ISBN 978-981-99-0887-5 (eBook)
https://doi.org/10.1007/978-981-99-0887-5

This Springer imprint is published by the registered company Springer Nature Singapore Pte Ltd.
The registered company address is: 152 Beach Road, #21-01/04 Gateway East, Singapore 189721, Singapore

The workshop CAMAD'19 was organized to express our gratitude towards Professor Hari Vasudeo Sahasrabuddhe, HVS to one and all, a pioneering computer scientist in the niche domain of computational musicology, to celebrate his 76th birthday.

"By having computers play the roles of musicians and musicologists in experiments we can improve our understanding of music."
—HVS

Research in computational musicology had been initiated in the 1960s, around the time HVS joined IIT Kanpur, choosing teaching in Science and Engineering as his career. Research in computational modeling was costly in terms of time and money. Probably due to these constraints, the domain was initially explored mostly by western scientists, and hence the research has been about western music. In 1992, Professor Sahasrabuddhe published his work on the analysis and synthesis of Hindustani classical music. Since then he has been a lighthouse for juniors like us who are venturing the sail through the ocean of musicology...

Infosys founder N R Narayana Murthy said, "The advice to take up learning over salary given by Professor Sahasrabuddhe helped me in choosing the right career path". In much the same way, HVS has been instrumental in shaping the careers of many students. We, on behalf of all of his students, dedicate this volume to Professor Sahasrabuddhe.
—Team CAMAD'19

Foreword

Indian performing arts have a glorious tradition with its roots going back over 2000 years. Indian classical music (ICM) as well as classical dance are highly evolved and stylized art forms with certain distinct features. ICM is a unique blend of formal apparatus—the "shastra", and freeform innovation, permitting performing artists to improvise extensively during a performance. There is a highly developed system of Ragas and Taals.

The performance involves an intricate interplay of melody and rhythm, almost a dialogue between these two. Musical compositions and ragas themselves are classified by their moods (bhaav), seasons, time of the day, etc. Within this theoretical apparatus, musical forms have evolved into gharanas which embody unique musical styles whose purity and sanctity are closely managed by the masters of the gharana. Musical education is mostly an oral tradition with long years of practical grooming under the watchful guidance of the teachers. Much of the music performed is not scripted.

Similar features can also be observed in dance forms like Bharatanatyam.

Modern electronics and digital signal processing have given us the capabilities of recording, transforming and communicating music. Using advanced digital signal processing techniques, much of today's popular music is deployed electronically and post-processed in studios enhancing their tonal characteristics. Electro-acoustic compositions made out of artificial, i.e., computer-generated sounds, are also prevalent.

The advent of Artificial intelligence gives us new capabilities of analysing, understanding and even generating multi-media content using computers. It provides us with an opportunity to approach the highly elaborate but informal and semi-formal knowledge embedded in artistic and multi-media content. AI with its novel ability to learn and extract knowledge from unstructured and semi-structured multi-media data provides a unique new ability to bring out the features of epistemic and linguistic content from visual and auditory data. It enables the musicology of Indian classical performing arts to be studied in a scientific way. It has the potential of transforming the pedagogy of music learning and appreciation.

The study of music and dance using scientific and computational techniques has grown over the last 30 years, especially for Western music. However, this is relatively unexplored in Indian classical music and dance. The COMAD'19 workshop fills the much-needed gap. Drawing upon the papers from the COMAD'19 workshop, the current volume presents forays into some seminal areas of computer analysis of Indian performing arts.

It is highly appropriate that the volume is dedicated to Professor Hari Sahasrabuddhe, a doyen of computer science and a pioneer who looked at the computational analysis of Hindustani music early on when such inquiries were almost unknown in the country. Perhaps, all this came naturally to Hari, with his background in Computer Science and with the influence of his wife, Mrs. Veena Sahasrabuddhe, a celebrated exponent of Gwalior Gharana. As a student at IIT Kanpur, I recall the marked influence the Sahasrabuddhe couple had on the musical scene of this elite technical institute. A cursory examination of Professor Sahasrabuddhe's google scholar page shows the many directions that he explored—these include exploration of the connection between music and bhaav, music similarity measures, musical information retrieval as well as broader explorations on the foundations like Raga modelling and Shrutis (musical scale). Professor Sahasrabuddhhe's work has left a lasting mark, inspiring many. In his retirement, Professor Sahasrabuddhe continues contributing and giving direction to the field. The current volume contains a keynote address as well as a technical paper by him. Clearly, this all is a labour of love for him.

The volume itself is a wealth of interesting papers, spanning various directions. These are categorised into "Computer assisted musicology", "Machine learning approaches to music", "Composition and choreography" and "Interfacing the traditional with the modern", and include works by leading researchers in the area. The editors, who are themselves established researchers, must be congratulated for putting together the excellent collection and also for their notable perspective.

I am confident that this volume will prove to be a valuable resource for the emerging field of computational analysis of Indian music and dance. I also take this opportunity to wish Professor Sahasrabuddhe many happy years of healthy and joyous life.

Professor Paritosh Pandya
TIFR
Mumbai, India

Preface

The Natyashastra (dramatics in Sanskrit), the 2500-year-old surviving Indian compendium on theatrical art, defined music as having three forms: vocal, instrumental and dance. It has referred explicitly to music in half of its chapters. Though music and dramatics have evolved along with society, the fact remains unchanged that they still share the traditional framework to a great extent. An international workshop, "Computer Assisted Music and Dramatics: Possibilities and Challenges (CAMAD'19)" held on February 25–27, 2019 in the Department of Computer Science, University of Mumbai, focuses on this definition.

AI has been driving traditional human centred activities to partly or completely automated processes. Automation has been attempted for a range of tasks, from identification and composition to judgement and appreciation of art performances, which have been considered the exclusive forte of human intelligence. The emphasis of CAMAD'19 was on developing computational models, as far as possible, to study music and allied fields. The selected 15 papers of the workshop are getting published under Springer's book series Advances in Intelligent Systems and Computing.

The first book in the Springer series on computational music science is from 2010. Soon enough the fourth title, Computational Musicology in Hindustani Music, was published in 2016. Though not part of the same series, we are privileged to publish the second book on computational musicology in Indian classical music by Springer. The 2016 book illustrates fundamental aspects like the role of statistics and introduction to a computational research platform, while in the present volume, readers will find the application of AI and ML to musicology. Topics like structural analyses, entropy, comparison of ragas and machine-assisted composition of music are thrust areas even now, and they will be researched in the future as well. Here they have been extended to the domains of instrumental music and dance. The therapeutic use of Indian classical music had been touched upon in the concluding chapter of the earlier book. This topic is important for society. For want of substantive objective proof, we could not consider two papers in this domain that were presented in CAMAD'19, including one by Guru Prem Vasantji.

Here, a set of 56 songs analysed by employing a regression-based learning model is shown to have captured the mood of a song with an accuracy of 73%. This study emerged out of a keynote speech, "Role of Prosody in Music Meaning," proposing features like tempo, melody and instrumentation to delve into classical music. A paper on music composition using PSO, and another on automating Bharatanatyam choreography, propose to direct and evaluate processes in accordance with the traditional framework without human intervention. These initiatives are worth pursuing. Exploring the Bharatnatyam Margam using graph theory has been considered for the first time. The paper on developing a musicality scale provides an elaborate computational process for checking a poetic piece for its potential to become a song. Such research is opening up new areas for computational exploration to heighten Indian music and dance forms in tune with the forthcoming technoscience era. They may attract the young generation. Yet another keynote speech "Epistemology of Intonation" was on the theoretical foundations of music, the computation of 22 shrutis and an interpretation of how they are instrumental in creating musical mood. An attempt to chart out directions for advancing computational musicology in the Indian context, by linking past and present knowledge and by employing associated technology, is made in "Computational Indian Musicology: Challenges and New Horizons". Such information presented by practicing stalwarts will develop deeper insights into the subject.

The distribution and ordering of the papers in suitable sections were shaped by the review process. The three papers in the first category entitled Computer Assisted Musicology address the extraction of musicological information from concert performances by processing their sound signals with the help of known techniques. Their outcome may yield feedback for future performances of seasoned musicians or could be employed in training novice learners. The five papers in the second category, Machine Learning Approaches to Music, are about the application of machine learning for processing metadata to extract pragmatic information. Research on the computation of aesthetics has evolved as an extension of the research on automating Bharatanatyam steps and sequences. These two papers, along with one on the automatic creation of tabla compositions, form the third category entitled Composition and Choreography. The fourth category is Interfacing the Traditional with the Modern. The papers here are suggestive of future trends. New perceptions of traditionally available information have been put forth. For example, music production has become an attractive career path for music-loving engineers.

A review paper on the evolution of research in musicology and another on the industrial applications of musicology research, and two more on the automatic classification and clustering of ragas from CAMAD'19, are not included in this book since the authors could not submit the final copy. In the context of non-Indian theatre, there are citations in the literature to the computation of the ontology of a play for purposes of action analysis, modelling suspense and dramatic arc in order to predict the success potential of a story, as well as the estimation of dramatic tension as a function of goals, obstacles and side effects. However, no paper on computational dramatics was submitted to CAMAD'19, though we had hoped to hear about research on the computational aspects mentioned in the Natyashaastra. Kathak dance guru Rajashree Shirke and acclaimed Marathi folk music performer Ganesh Chandanshive, fascinated as they were by the idea of applying computation to their domains of the performing arts, presented computable ideas in their respective domains. In the absence of implementations, at this stage, we are not able to include these papers in these Proceedings, as also another paper related to dramatics. We look forward to collaborations between the authors of these papers and interested researchers in the domains of music, computer science, mathematics and cognitive science to take these ideas to their logical conclusion.

As many as 18 authors have contributed papers to this volume. Interestingly, about a third of them are professional musicians, while the remaining are AI or computer scientists, about half of whom are formally educated in Indian classical music. For about a third of the authors, these papers are part or extensions of their doctoral research. In a way, this mix testifies to the variety and quality of the content.

Mainly because of circumstances due to the pandemic, this publication project took almost three years for its completion. Substantiating the writings with statistically proven results was a major task. This kind of experience is no less than that of completing a doctoral thesis. It is a matter of professional satisfaction that all the authors took the observations and suggestions in the right spirit and willingly revised their drafts, sometimes more than once. Our thanks are due to all of them for being with us, patiently and passionately, throughout this journey. Special thanks are due to Professor Hari Sahasrabuddhe for guiding us all through and Professor Jayant Kirtane for editing all the drafts for enhancing their readability. This project would not have been completed without the consistent support of Professor Vivek Patkar who critically went through each and every draft and provided clear and constructive feedback. What to say about Mr. Srijan Deshpande? After completing his own paper, he volunteered to help us organise the material under four categories from the practitioners' perspective. We thank the authorities of the University of Mumbai, the then Pro Vice Chancellor Professor R D Kulkarni, in particular, for offering grants to host an array of world-class speakers at CAMAD'19. Thanks to our family and friends at our homes and our professional homes for providing all the required support. We have no words to express our gratitude and appreciation for the patience, consideration and guidance that we experienced at Springer. They agreed to produce this volume without expecting any financial support from our side.

We are sure that this book about the advances in computing and musicology specific to Indian music will receive due attention from researchers across the globe with varying perspectives. The proceedings should help discover fresh avenues and break new grounds to cater to emerging tastes and exploit advanced technologies. The readers of this volume would expect the next one. *Ye dil mange more* … .

Mumbai, India — Ambuja Salgaonkar
Pune, India — Makarand Velankar

Contents

About the Editors

Ambuja Salgaonkar has a Ph.D. in Computer Science, an M.B.A. (Operations Mgmt), an M.A. (English Lit), and about 30 years of experience in teaching at the university level and research in problem-solving using AI. She has extensively researched Indian heritage science for contemporary applications, including Katapayadi numbers and Shree yantra type of designs for information retrieval and security, exploring the syntax of the Indus script as a consistent messaging system and image processing of palm leaf manuscripts. Her recent published work is in motion planning for agricultural robots, automatic question generation, and Konkani–Hindi machine translation. She has been coordinating corpus-generating activities related to Marathi and its dialects for the Bhashini project of the Government of India. She is the Marathi translator of a collection of 49 essays with the title "India's cultural history up to 1947 CE," a multi-lingual project with international collaborators. Ambuja has been a student of Indian classical music (violin). She has been fortunate to receive guidance from Professor Hari Sahasrabuddhe for her various researches including her Ph.D. work. She is a prolific writer in Marathi. Her articles on Vidushi Veena Sahasrabuddhe, Vidushi Sushilarani Patel, and Dr. Anjali Nigwekar have been well received. Transcreation of *Alice in Wonderland*, translation of Tagore's *Geetanjali,* and Haiku forms of Kabir's Dohas are her contributions. She has been a recipient of Adya Marathi Haikukar Srimati Shirish Pai award for her contributions in Tipedi, a collection of her investigations and demonstrations in novel Marathi haikus. Ambuja's current passion is educational technology. She was instrumental in designing as many as ten courses to teach Indian classical music in distance and open learning mode. She successfully conducted a three-semester specialization in computer-assisted music learning at the University of Mumbai. Creation of a specialized MOOC on conjoining Ravindra Sangeet with Hindustani classical music and developing a scale for measuring the complexity of composition are her dream projects.

Makarand Velankar M.E., Ph.D. in Computer Engineering from SP Pune University, has about 11 years of industry experience. Later, he joined MKSSS's Cummins College of Engineering for Women, Pune, and has been teaching there for the last

21 years. His passion for research in computational musicology, developed through interactions with Professor Sahasrabuddhe, has led him to explore the world music canvas with a focus on Indian music. His doctoral research on query by humming, content-based retrieval, modelling melodic similarity, sentiment analysis, performance evaluation, and ML-based recommendation systems has received appreciation in conferences like ISMIR. His work in this domain has been published in reputed journals. Developing a commercially available personalized music recommendation system is his immediate goal. In recent times, he has been engaged in exploring the domain of automatic generation of music.

Makarand's passion for entrepreneurship led him to become a start-up mentor for Wadhwani AI, a multinational NGO located in Mumbai. So far, he has mentored more than five student start-ups and provided consultancy to two established business setups to scale up. He has been heading a pre-incubation center at his college. He has also initiated and nurtured a music technology group.

Computer Assisted Musicology

Bridging the Gap Between Musicological Knowledge and Performance Practice with Audio MIR

Preeti Rao

Introduction

Classical music, also termed art music, is considered to be a highly aesthetic form rooted in a specified theoretical framework. It also typically implies the availability of written notation. Much of the research in musicology (i.e. the academic or scientific study of music) of Western classical music has involved the written form of music known as the "score". Only recently, have performance aspects linked to interpretation and expressiveness in the rendering of the pre-composed music begun to gain attention. In this case, the audio recording is the basis of the study of what are essentially considered departures (in rhythm and dynamics) from the notated score. On the other hand, among the popular practices employed by Western musicologists to study non-Western repertoire, has been the repeated playback of recordings to achieve some form of transcription. A piece of music can thus be analysed for musical traits such as tempo, phrase length and types of pitch movements [1]. In contrast to this laborious and somewhat subjective analysis, the direct measurement of the physical sound such as achieved by Seeger's sonograph, an electronic transcription system, was viewed as a means to increase the scope and accuracy of empirical studies [2]. On similar lines, the field of computational (or digital) musicology seeks to use visual or statistical representations computable from the physical audio signal that can be then applied to compare pieces of music.

Indian classical music, also considered an old and sophisticated tradition, is based entirely on oral transmission and makes very little use of written notation. Both the major genres of Indian classical music, North Indian (Hindustani) and South Indian (Carnatic), are associated with a theory and pedagogical practices that have remained relatively unchanged over several decades. With origins in folk music, both genres

P. Rao (✉)
Department of Electrical Engineering, IIT Bombay, Mumbai, India
e-mail: prao@bhairav.ee.iitb.ac.in

A. Salgaonkar and M. Velankar (eds.), *Computer Assisted Music and Dramatics*, Advances in Intelligent Systems and Computing 1444,
https://doi.org/10.1007/978-981-99-0887-5_1

have evolved into highly structured performance forms based on theory, also known as the raga grammar [3, 4]. Given the current vibrant classical music performance scenario in the country, as well as the legion of great artists who are well known through their concert recordings carried out over the past six decades or so, we can expect that we have a lot of audio data that musicology research stands to benefit from. Some interesting questions that can potentially be answered based on the analyses of audio recordings are around the differences in performance style between different schools (gharanas), different artists, and across instrumental and vocal forms. In this paper, we consider the application of computational representations to the description of the melodic aspects of performance with reference to the underlying raga grammar. We expect such a study to provide insights about performance practice that are not evident from the theory alone. In the next section, we present a description of raga grammar. This is followed by a review of signal processing methods applied to audio recordings to obtain melody representations. Finally, we present examples of applying computational representation to the study of performances.

Raga Grammar

The permitted notes or 'svaras' constitute the tonal material of a raga, while their relative importances constitutes the tonal hierarchy. These descriptions, known as distributional properties, are also used in the context of key belongingness in Western music [5]. Further, in raga music, also specified are the svara sequences (aroh, avroh and the characteristic phrases). The svaras are to be interpreted as pitch intervals relative to the tonic chosen by the artist. The phrases are sequences of the svaras, where a raga svara now appears in specific contexts of the preceding and succeeding notes. A raga can comprise as few as five notes (pentatonic) or include over seven notes. The notes are assigned solfege labels but it is important to keep in mind that the precise intonation of a given note (with the same label) depends on the raga and is part of the implicit knowledge acquired about the raga grammar in the course of music training. The *vadi* and *samvadi* specify the most prominent (dominant) and second most prominent (sub-dominant) notes, respectively. While prominence is associated with musical emphasis (just as in spoken language), it is not clear how this attribute is meant to be realized in practice. Is the more prominent note the one used most frequently, or is it expected to be longer in duration in each occurrence, or does it coincide with specific metrical accents of the *tala* (rhythmic cycle)?

Figure 1 provides a typical definition of raga grammar, as compiled from musicology texts [6, 7–9], in a form that a teacher might narrate to a student. In order to interpret it better, we present the comparison of the grammar of two ragas that are considered to share several characteristics. The pentatonic ragas Deshkar and Bhupali are known as "allied ragas". They have the same set of svaras (S; R; G; P; D corresponding to 0, 200, 400, 700 and 900 cents, respectively) and common phrases in terms of svara sequences. The svaras and the hierarchy in terms of vadi and samvadi notes constitute the distributional representation of the raga. The acoustic manifestation of the hierarchy of svara can be studied by signal analyses. The last row in Fig. 1 points to the precise intonation of the svaras and is also a part of the distributional representation. We may interpret it as telling us that the R, G and D notes of Bhupali are just intoned intervals while the same svara in Deshkar are realized with slightly higher intonation. Note that the specification of interval size is not precise. This is an example of a phenomenon that can be studied via measurements of the physical sound in the music recordings. The aroha (ascent), avroha (descent) and the phrases constitute the structural representation. They indicate the basic building blocks of the melody. Improvisation is a strong trait of Indian classical music but one that is within the specified structural constraints. Thus, while the artist is free to compose the melody in the moment during a performance, the svaras necessarily come from the tonal material and the phrases that are characteristic of the raga. Other structural constraints include the local tempo and the overall timing provided by the rhythmic cycle of the tabla (percussion) known as the *theka*, principally the cycle boundaries which are marked melodically by the refrain (mukhda) of the chosen composition [10].

We see from Fig. 1 that both ragas share several phrases. Deshkar has the parenthesized versions of R, D and P in certain svara contexts. These correspond to the *alapa* form or non-emphasized form. So, the same svara sequence GRS would be rendered as G(R)S in Deshkar implying a difference in the melodic shape of the motif where it is de-emphasized (i.e. shortened) with respect to the neighbouring notes. Once again, this is an interesting aspect that can be validated, as well as more precisely described, via signal measurements. In the present work, we apply computational methods to a dataset of vocal concerts by well-known artists of Hindustani classical music.

Computational representations of the melody can thus help us interpret the distributional and structural specifications, prescribed by the theory, far more precisely via measurements of the actual acoustic realizations of the svaras and phrases, which

Deshkar	Bhupali
Tonal material: SRGPD	Tonal material: SRGPD
Ar: SGPD, SPDS *Av*: S, PDGP, DPG(R)S	*Ar*: SRG, PDS *Av*: SDP, GDP, GRS
Vadi: D, *Samvadi*: G	*Vadi*: G, *Samvadi*: D
Phrases: SG, G(P)DPD, P(D)SP, DGP, DPG(R)S	Phrases: RDS, RPG, PDS, SDP, GDP, GRS
Higher shrutis of R, G, D	Natural shrutis of R, G, D

Fig. 1 Specification of raga grammar for the two allied ragas Deshkar and Bhupali under study

presumably have been learned by the artists implicitly in the course of their training. Also interesting to capture, is the extent of variability, if any, in the svara duration and intonation or melodic shape of a given phrase over the course of the concert or across concerts and artists. Finally, computational representations can help us compare two performances of the same raga to appreciate the nature of improvisation. Improvisation involves sequencing the ‘building blocks’ of the raga, namely the svaras and the phrases or motifs, in different ways as the concert progresses in time [11]. Thus. two concerts in a given raga by the same artist at two different times can, for instance, exhibit different melodic progression patterns in keeping with the tonal distributional properties overall and the structural constraints of the rhythm cycle [12, 13]. In the next two sections, we present the audio signal processing and representation methods that facilitate the proposed musicological investigations.

Audio Signal Processing

Music signals are periodic and elicit a perception of pitch linked to the fundamental frequency of the tone. The remaining attributes of the physical signal corresponding to a single note are its spectral envelope and intensity. The timbre (related to instrument or voice identity) is largely captured by the spectral envelope. A Hindustani vocal concert has a single predominant melodic voice (sometimes accompanied by a second melodic voice such as the harmonium or sarangi) with the tabla providing rhythmic accompaniment and the tanpura, the drone. Audio MIR is a field that extracts semantic information from audio signals by linking low-level signal properties such as the fundamental frequency, event onsets and spectral envelope with high-level music attributes such as melody, rhythm and timbre. Similarity, which forms the cornerstone of music retrieval, is then defined in terms of a distance measure computed between representations based on the high-level music attributes.

Figure 2 shows the processing pipeline for the computation of a melody representation from the audio signal or waveform of a piece of music [13]. The processing is suited to comparisons of the tonal content and tonal hierarchy across performances in the same and in different ragas. The fundamental frequency or pitch of the predominant voice is detected at 10 ms intervals (i.e. at the rate of 100 times/s) to obtain the vocal melodic contour in all previously detected singing voice containing regions [14]. Pitch detection algorithms based on the short-time spectrum can cluster harmonics corresponding to the different co-occurring instruments due to the sparsity of each source in the spectrum. We use a predominant pitch detection algorithm that exploits the spectral characteristics of the singing voice coupled with analysis settings that take into account the singer's pitch range and singing style [15]. The analysis settings are primarily the short-time spectrum computation window, the pitch search range and the temporal smoothness constraint used in tracking the time-varying pitch. Appropriate analysis presets depend on the singer's gender and the singing style in terms of the speed of variation, and facilitate accurate pitch racking of the vocals as long as the singer's voice is clearly audible above the accompaniment.

The tonic is separately extracted by one of several available tonic detection methods [16]. Some of these use the drone regions in the singer's pause regions, while others obtain it via multipitch analysis. The pitch contour is interpolated over the very short unvoiced regions corresponding to the consonants in the singing. We thus obtain a continuous pitch contour that, after tonic normalization, can be viewed as a complete representation of the melody of the piece, one with implicit information about the tonal content and hierarchy as well.

A pitch histogram can be computed from the pitch contour samples to obtain a pitch salience histogram. A fine bin width of one cent can provide a smooth histogram over the range in which the svara locations are seen as clear peaks in Fig. 2. We note that this histogram is influenced by the note transitions and ornaments. The relative heights of the peaks signify the occurrence frequency of each narrow range of pitch values. The highest peak can therefore be assumed to correspond to the dominant note of the raga. The remaining local peaks should correspond to the raga svara locations. The pitch salience histogram provides a detailed description of the distributional characteristics in terms of the pitch interval location and extent of occurrence of

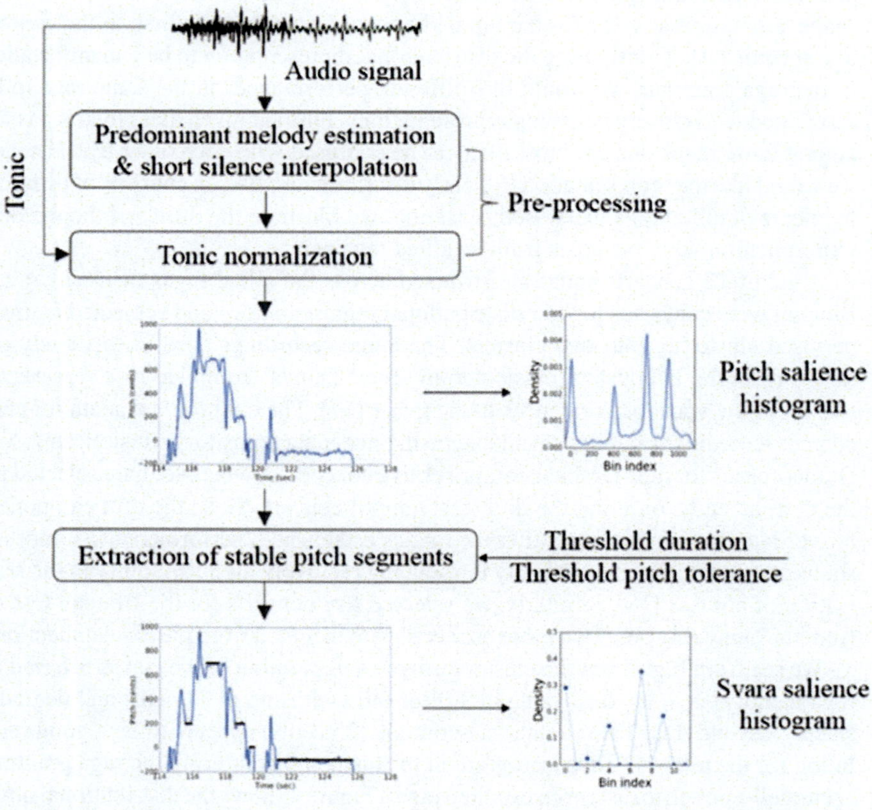

Fig. 2 Block diagram of the signal processing from audio signal to pitch distributions [13]

the svara, far beyond that captured by the raga grammar statements. An alternate distribution is a discrete distribution resulting from retaining only the stable note regions obtained by picking segments above a specified duration in the contour that is localized in pitch to a neighbourhood of the svara pitches (identified by the locations of the peaks in the pitch salience histogram). The svara salience histogram is a compact representation of the tonal hierarchy. The discrete distribution is similar to the 12-bin histogram used to represent the tonal hierarchy in Western music, known as the pitch-class profile, which has formed the basis for key detection algorithms [5].

Distributional and Structural Representations

The pitch histogram represents distributional information. We note that a number of analysis parameter settings are required which influence the visual representation and consequently its effectiveness in a given MIR task. The bin width is an important parameter that changes the smoothness of the distribution and consequently the shape. An optimum value for the bin width would have to be defined in the context of a specific task. Considering the distributional representation to be a manifestation of the raga grammar, we would like different performances in the same raga to be associated with closely matching representations. Further, given that allied ragas are among those ragas that are most likely to be confused with each other by a listener, we would like the representation to clearly discriminate between concerts of different members of allied raga pairs. In this section, we illustrate the utility of the melodic representation with examples from an allied raga pair.

A set of 12 concerts equally distributed across the allied ragas of Deshkar and Bhupali were converted to their distributional representations and subjected to unsupervised clustering into two clusters. The audio recordings used in this study are drawn from the Hindustani music corpus from 'Dunya' compiled as a representative set of the vocal performances in the genre [17]. The editorial metadata for each concert recording is publicly available on the metadata repository MusicBrainz. The Dunya corpus for raga Deshkar comprises five concerts of which four are selected for the current study, omitting the drut (fast tempo) concert. Since the drut component arrives late in a concert, well after the raga is established, performers use it more to showcase their technical virtuosity introducing relatively high variability in the realization of phrases [18]. Similarly, we selected five concerts for the Bhupali test set from the Dunya corpus. Two more concerts were included from personal collections.

We see from Fig. 3 that the cluster purity (i.e. separation of the concerts based on raga identity) is at its ideal value of 1.0 for bin widths up to 27 cents and degrades steeply beyond this. Thus, a quarter-semitone (25 cents) appears to be a good resolution for the pitch salience histogram in the context of capturing the raga grammar accurately enough to discriminate allied ragas. Figure 4 shows the distributional information (pitch interval locations and relative strengths) captured by each of the continuous and discrete pitch histograms. The latter, termed the svara (or note) salience

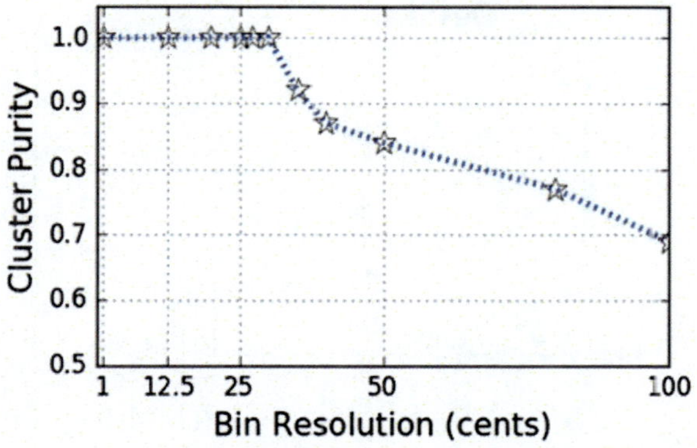

Fig. 3 Clustering performance for different bin widths

histogram, is computed from the stable note segments of the melodic contour, omitting the transition regions. We observe in Fig. 4 that while peaks corresponding to the svaras are similarly located in the two ragas, the relative heights of the peaks follow distinctly different patterns, consistent with the tonal hierarchy of each of the ragas. In terms of cluster purity, the svara salience histogram gives a relatively high value of 0.96 indicating that the segmented stable notes capture the tonal hierarchy nearly as well as the entire continuous melodic contour, at least on the time scale of the full concert [13].

Either of the distributions (continuous or discrete) depicted in Fig. 4 can serve as a template for the classification of performances based on raga. We would need to define a distance measure to compare two representations. The choice of a specific model and the values of its parameters can be optimized based on data in the form of concerts labelled by raga. A variety of distance measures was considered in the context of discriminating concerts drawn from allied pairs of ragas to find that the Bhattacharya distance between distributions performed the best [13]. The Bhattacharya distance is a popular statistical measure of the similarity between two distributions [19].

The structural representation of a concert, on the other hand, comprises the characteristic phrases of the raga. The melodic pitch contour can be viewed as a sequence of phrases and intervening notes and transitions. Figure 5 shows an extract of the pitch contour from a concert in the raga Alhaiya-bilawal superposed with a musician's transcription. We see how the transcription comprises phrases rather than separated svara, the phrases being recognizable gestalts from the melodic shapes. Certain key features of a phrase are the relative durations of the stable svara regions and the transitions linking these. Based on such features, it is possible to automatically segment specific phrases from the melodic contour provided, of course, we take into account the expected variability in the melodic shape arising from a phrase's context [20].

In our dataset of 12 concerts in the allied raga pair of Deshkar and Bhupali, we annotated a single phrase 'GRS' common to the two ragas, and measured the duration and intonation of each landmark event within it. A notable difference between the tonal distributions in Fig. 4 is the strength of svara R, which is relatively low in raga

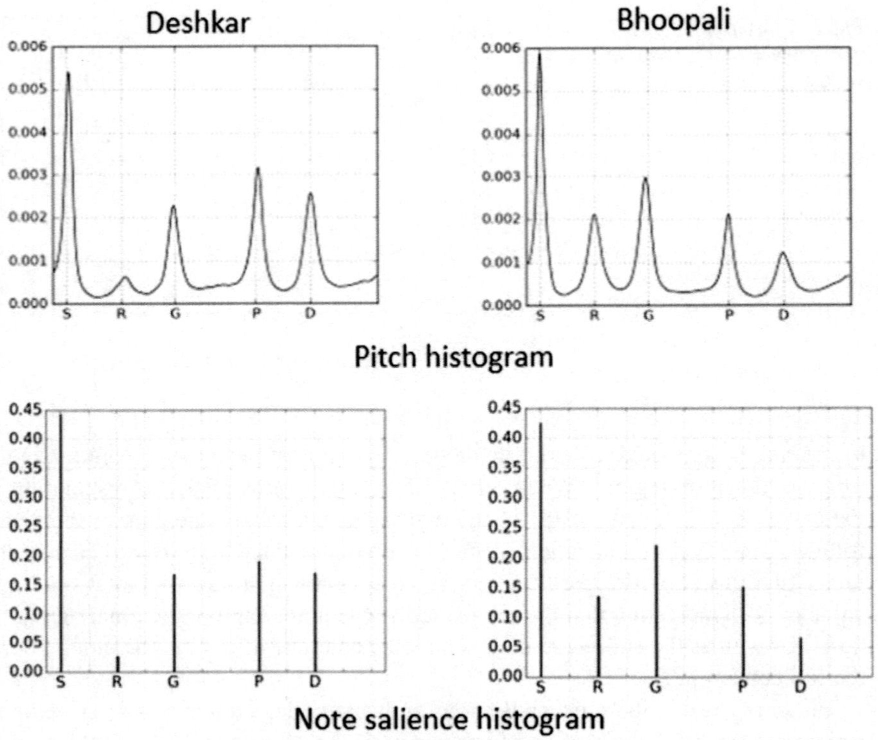

Fig. 4 Pitch salience histogram (top) and svara salience histogram (bottom) for one concert of each raga in the allied raga pair

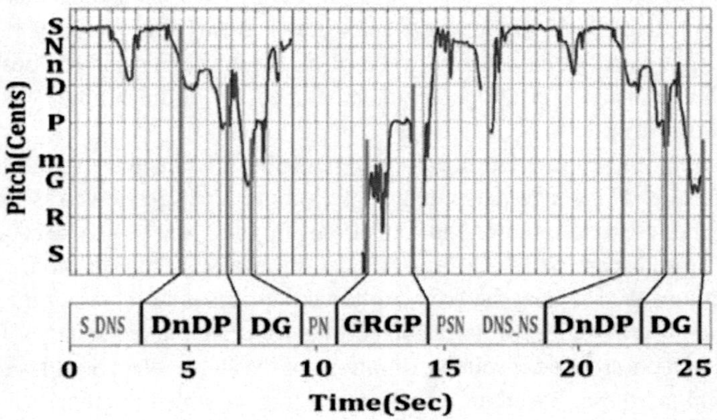

Fig. 5 Pitch contour of a 25 s extract with manual transcription by a musician

Deshkar. This aspect is expressed by the parentheses around R in the mandatory raga grammar (Fig. 1); it is an aspect well known to musicians in that R is visited but not held in raga Deshkar, while it is not particularly constrained in raga Bhupali. Box plots of the different measurements carried out of the individual events (svaras and glides) of the GRS phrase are shown in Fig. 6 for each of the two ragas across 12 concerts. We again note the sharp contrast in the distributions of the svara R, as well as in the durations of the transitions to and from R. This suggests that there is considerable flexibility in the durations (possibly constrained only by the local context) of all the notes except for R. The latter is carefully realized in a specified absolute duration in different concerts by different artists and forms a distinctive feature of Deshkar. Box plots of the measured pitch intervals of the note G, also shown in Fig. 6, illustrate the distinct intonations of the same svara in the two ragas, consistent with the theory. The computational representation thus helps us interpret the theory more precisely in terms of the size of the difference of pitch. Recent work on the perception of synthesized phrase shapes validated the critical role of the distinguishing feature of R duration in the observed categorical perception phenomenon with musicians trained in the genre [21].

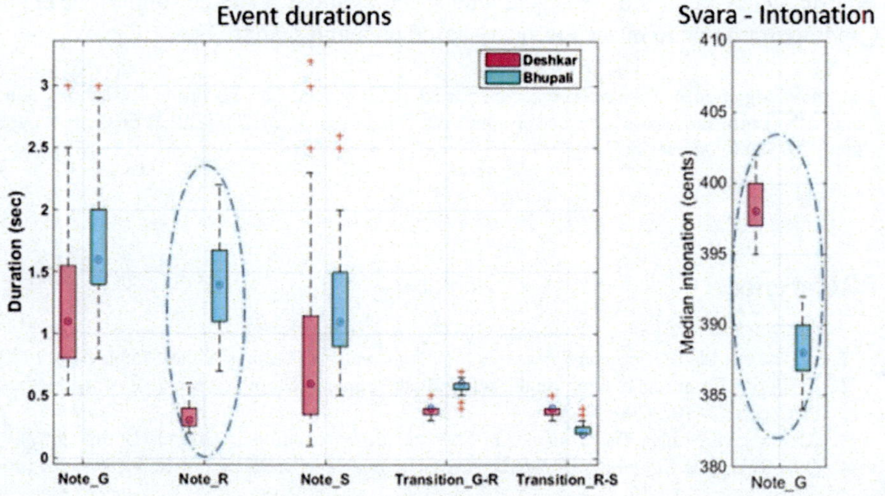

Fig. 6 Distributions of 'event' *durations* and intonations of the note G across the annotated GRS phrase instances in the six concerts each in the two ragas Deshkar and Bhupali [22]. The distinctions mentioned in the theory are encircled

Conclusion

We have demonstrated how computational methods can be useful in musicology research on performance practice. Automatic processing can contribute to greatly increasing the scope of studies in genres such as Indian art music. An example was presented of using acoustic features related to the melody derived from concert recordings to obtain deeper insights into how raga grammar constraints are manifested in performance. Further, the models help us discover systematic practices within and across performances that are not explicitly verbalized in the course of pedagogy. The model can also help us to abstract a performance into the invariant theoretical constructs and the more variable improvisational aspects. This can be applied to the critical analysis of performance and is of value to musicology studies. Other potentially interesting investigations concern the dependence of musical attributes such as style on the period and gharana. With the internet contributing to the globalization of society, cultural artifacts such as the music of a specific region are increasingly accessible across the world providing artists a window to new audiences. Such opportunities for exposure of audiences to diverse musical styles and the associated increase in their breadth of musical knowledge can be facilitated by the outcomes of such musicology research. Finally, considering applications in digital music technology, similarity is the foundation of music recommendation systems with melodic similarity being an important component of it. In the case of raga music, where ragas are associated with specific moods, melodic similarity can play an important role in music discovery based on audio search.

Acknowledgements This work received partial funding from the European Research Council under the European Union's Seventh Framework Programme (FP7/2007-2013)/ERC grant agreement 267583 (CompMusic).

References

1. W. van der Meer, *Hindustani Music in the 20th Century* (Martinus Nijhoff Publishers, 1980)
2. E. Clarke, Empirical methods in the study of performance. Empirical Musicol Aims, Methods, Prospects 77–102 (2004)
3. K.G. Vijaykrishnan, *The Grammar of Carnatic Music* (Mouton de Gruyter, Berlin, 2007)
4. D.S. Raja, *The Raga-ness of Ragas: Ragas beyond the Grammar* (D. K. Printworld, India, 2016)
5. C. L. Krumhansl, *Cognitive Foundations of Musical Pitch. Chapter 4: A Key-Finding Algorithm Based on Tonal Hierarchies* (Oxford University Press, New York, 1990), pp. 77–110
6. S. Rao, J. Bor, W. van der Meer, J. Harvey, *The Raga Guide: A Survey of 74 Hindustani Ragas* (Nimbus Records with Rotterdam Conservatory of Music, 1999)
7. Music in Motion: The automated transcription for Indian music (AUTRIM) project by NCPA and UvA. https://autrimncpa.wordpress.com/. Last accessed: 19 Sept 2017
8. Distinguishing between Similar Ragas. http://www.itcsra.org/Distinguishing-betweenSimilar-Ragas. Last accessed: 19 Sept 2017
9. V. Oak, 22 shruti. http://22shruti.com/. Last accessed: 19 Sept 2017

10. J.C. Ross, T.P. Vinutha, P. Rao, Detecting melodic motifs from audio for hindustani classical music, in *Proceedings of the 13th International Society for Music Information Retrieval Conference (ISMIR)* (Porto, Portugal, 2012)
11. L. Nooshin, R. Widdess, Improvisation in Iranian and Indian music. J. Indian Musicol. Soc. **36**, 104–119 (2006)
12. K.K. Ganguli, S. Gulati, X. Serra, P. Rao, Data-driven exploration of melodic structures in Hindustani music, in *Proceedings of the International Society for Music Information Retrieval (ISMIR), 2016,* New York, USA, pp. 605–611
13. K.K. Ganguli, P. Rao, On the distributional representation of ragas: experiments with allied raga-pairs. Trans. Int. Soc. Music Inform. Retrieval (TISMIR) **1**(1), 79–95 (2018)
14. V. Rao, P. Rao, Vocal melody extraction in the presence of pitched accompaniment in polyphonic music. IEEE Trans. Audio, Speech Lang. Process. **18**(8) (2010)
15. S. Pant, V. Rao, P. Rao, A melody detection user interface for polyphonic music, in *Proceedings of the National Conference on Communications (NCC)* (Chennai, India, 2010)
16. S. Gulati, A. Bellur, J. Salamon, H.G. Ranjani, V. Ishwar, H.A. Murthy, X. Serra, Automatic tonic identification in Indian art music: approaches and evaluation. J. New Music Res. (JNMR) **43**(1), 53–71 (2014)
17. X. Serra, Creating research corpora for the computational study of music: the case of the Compmusic project, in *Proceedings of the 53rd AES International Conference on Semantic Audio* (London, 2014)
18. S. Kulkarni, *ShyamraoGharana*, vol. 1. (Prism Books Pvt. Ltd., 2011)
19. T. Kailath, The divergence and Bhattacharyya distance measures in signal selection. IEEE Trans. Commun. Technol. **15**(1), 52–60 (1967)
20. K. K. Ganguli and P. Rao, A study of variability in raga motifs in performance contexts. J. New Music Res. 1–15 (Feb 2021)
21. K.K. Ganguli, P. Rao, On the perception of raga motifs by trained musicians. J. Acoust. Soc. Am. **145**(4), 2418–2434 (2019)
22. K. K. Ganguli, P. Rao, Towards computational modeling of the ungrammatical in a raga performance, in *Proceedings of the 18th International Society for Music Information Retrieval (ISMIR), 2017,* Suzhou, China

Spectral Analysis of Voice Training Practices in the Hindustani Khayāl

Srijan Deshpande

Introduction

Computer-aided musicology is certainly not a new phenomenon and significant work has been done, even in the Indian context in this domain. Perhaps a defining characteristic of the computational musicology of Indian music has been its choice of the classical music of India as its subject of study. This is unsurprising given that both the Hindustani and Carnatic genres of music have a substantially standardized textual theory which often functions as a useful starting point and reference for applying computational methods. Perhaps one of the chief reasons the classical genres of music are more likely to be studied using computational tools is their 'classical' status, which is in itself a product of the social as much as of the aesthetic history of these genres [1, 2]. A fallout of this is the fact that computational analysis of Indian music tends to focus primarily on the issues of *shrutī* (microtonality and intonation) and *rāg* identity, as is apparent from the survey of recent work in the field found in [3]. A primary motivation for these studies seems to be to explore the possibility of measuring performed music against standardized theory. These approaches are certainly useful in their ability to generate and make use of empirical data and have also opened up entirely new fields of study such as music information retrieval. Yet, it is uncommon to find research that uses computational methods to unpack information available in the *oral* and *performative* traditions of these kinds of music, rather than in their textual-theoretical traditions. It is this gap that the present study aims to address.

One aspect of performed Indian music that seems not to have been substantially addressed is that of musical timbre. Recent advances in the understanding of the

S. Deshpande (✉)
Manipal Centre for Humanities, MAHE, Manipal, India
e-mail: srijand@gmail.com

A. Salgaonkar and M. Velankar (eds.), *Computer Assisted Music and Dramatics*, Advances in Intelligent Systems and Computing 1444,
https://doi.org/10.1007/978-981-99-0887-5_2

human singing voice facilitated by the disciplines of voice science and psychoacoustics, as well as the easy availability of tools for spectrographic analysis have revealed that a vocalist's timbre is a storehouse of musicological information. A master musician imparting voice training to a disciple is an example of a situation where the question of timbre carries importance. This is a particularly potent situation that can reveal important details about the singer's stylistic, aesthetic and pedagogical goals, choices, and decisions. This paper will, then, demonstrate how it is possible to use contemporary research in voice science and psychoacoustics to explicate and problematize master musicians' ideas of voice training. We will look at a specific voice training strategy employed by the master twentieth-century vocalist Pandit Kumar Gandharva and attempt to develop a nuanced understanding of his timbral goals as demonstrated by his own performance, his pedagogical strategies, and his articulated ideas on the subject of training the singing voice.

Voice Science and Psychoacoustics: Brief Review of Recent Work

The current study depends largely upon the work of Ingo Titze, Kenneth Bozeman, and Ian Howell, each of whom has been instrumental in developing our understanding of vocal habilitation, vocal acoustics, and psychoacoustics, respectively. All three work primarily within the domain of voice pedagogy and are the leading scholars in the field. The following is a brief review of the theories propounded by these scholars that are immediately relevant to this study.

The Non-linear Source-Filter Model, Harmonics and Formants

The present study makes use of the non-linear source-filter theory of vocal acoustics. In essence, this theory postulates that vocalization consists of a sound pressure waveform that is created at the vocal folds (the source) and shaped by the vocal tract (the filter). The source generates a waveform that is rich in harmonics, which are all multiples (H2, H3, H4, etc.) of the lowest common denominator frequency called the fundamental frequency or the first harmonic (H1). The fundamental frequency is also alternatively denoted as F0, but because this paper also deals with formants which are denoted F1, F2, F3, etc., we will avoid potential confusion by using the notation H1 to refer to the fundamental frequency of the voice source and H2, H3 and so on to refer to the subsequent harmonics in the series.

These harmonics are selectively emphasized or deemphasized (filtered) by the vocal tract, depending upon its shape and size at the moment of phonation. Specifically, the length and shape of the vocal tract define the locations of its *formants.*

Formants are specific frequency ranges at which the various air columns in the vocal tract resonate, and are denoted F1, F2, F3, and so on. In general, harmonics that are close to formant frequencies receive a boost in intensity, while those that are further away from the formant frequencies lose intensity. The perceived timbre of any singing voice is thus a composite sound that is made up of and defined by the relative intensities of its component harmonics. Because the vocal tract, the resonator of the human vocal instrument, is unique in its ability to change size and shape, vocalists continually alter these parameters, thereby changing the intensities of the harmonics in their sound signal, giving them a vast and varied palette of timbral options to choose from [4].

The term 'non-linear' refers to recent advancements in our understanding of this model. In certain circumstances such as when singing with an occluded or semi-occluded vocal tract, as discussed in the case study below, 'acoustic energy passing through the filter can be productively reflected back onto the source, assisting the efficiency and power of the voice source/vibrator' [4]. The harmonic spectrum thus generated can be analyzed using spectrographic tools and interpreted fruitfully with reference to a musicological understanding of the sound being analyzed. In the present study, such analysis has been done using the VoceVista Video tool [5].

Interpreting the Spectral Envelope

Roughness and Resolvability

Ian Howell's work [6, 7] on bringing knowledge from psychoacoustics to vocal pedagogy has given us a number of insights into the timbre of the singing voice. Prominent among these are the concepts of 'Roughness and Resolvability' and 'Absolute Spectral Tone Color' (ASTC). Howell describes *Auditory roughness* as the perception of a buzzing quality that arises because the cochlea cannot differentiate between simple tones 'that are very close in frequency' [7]. Howell also shows that tones that are separated by an interval of a minor third or less will give rise to such roughness. Since all harmonics of a voice beginning from H5 onwards satisfy this criterion, they contribute roughness to a singer's timbre, and this perception of roughness is directly proportional to the strength of higher harmonics [7]. Additionally, Howell builds upon the well-established acoustics concepts of the missing fundamental to show that only the first eight harmonics neatly resolve into the fundamental pitch, while higher harmonics do not. This leads him to the conclusion that for harmonics higher than H9, each successive harmonic appears to be 'part of a separate percept' rather than of the fundamental frequency. Both these concepts are crucial to the discussion presented here. The following Table 1 summarizes this information.

Table 1 Roughness and resolvability. Based on observations in [7]

Harmonic Number	H1	H2	H3	H4	H5	H6	H7	H8	H9	H10	H11	Hn
Pitch resolvability	Resolved								Unresolved			
Roughness	Pure				Rough, progressively rougher							
Summary	Pure and resolved				Rough and resolved				Rough and unresolved			

Absolute Spectral Tone Color (ASTC)

Howell has also brought to the field of vocal acoustics the concept of Absolute Spectral Tone Color or ASTC. Essentially, ASTC theory tells us that the human ear ascribes particular vowel-like qualities to certain pitches, *irrespective of the source of the sound.* For instance, a simple sine tone of around 1000 Hz, will inevitably be perceived as possessing an /ɑ/ vowel-like quality. The following Table 2 summarizes the ASTC vowel qualities (denoted using their IPA symbols) that various frequencies possess.

Table 2 shows that these frequencies/vowel qualities also carry with them connotations of acoustic 'brightness'. Thus, lower frequencies and their associated vowels are perceived as 'dark', and vice versa.

Note: This table is an approximate summary put together for the purposes of this paper, based on [6]. Note that there are overlaps in the frequency ranges depicted below. In the actual perception, the perceived vowel gradually transitions from an /u/ quality to an /i/ quality as the frequency increases. [8] is a revealing demonstration of both ASTC and auditory roughness.

Armed with this knowledge and with the tools to apply it to specific cases, it now becomes possible for us to conduct spectral analyses of samples of the singing voices of master musicians and identify their particular timbral goals, as well as to hypothesize about possible reasons behind their choices.

Table 2 Absolute spectral tone color. Based on observations in [7]

Frequency Range (Hz)	Upto 450	450-750	750-1000	1000-1300	1300-1500	1500-1900	1900-2300	2300-3500	3500 Onwards
Perceived Vowel Quality (IPA)	u	o	ɔ	ɑ	a	ɛ	e	i	Bright i
Perceived Brightness	Dark								Bright

Case Study: Examining Aspects of Kumar Gandharva's Pedagogy of the Voice

One of the things Kumar Gandharva was most known for was his mastery over intonation—he is recognized as one of the most tuneful vocalists in the Hindustani tradition and is often compared with the likes of Abdul Karim Khan, another master of intonation [9]. Commentators like Deshpande even go so far as to say that Gandharva's experimentation with vocal timbre was one of the defining features of his music [10]. While these facts are enough to justify studying his timbral goals, Gandharva was also particular about his disciples cultivating a good singing voice and had formulated definite pedagogical strategies which he articulated in detail in his available interviews. The interview considered here is perhaps his most detailed exposition on the subject and is available as an audio recording as well as in print [11, 12]. This presents us with the opportunity to conduct actual acoustical analyses of his demonstrations and correlate them with his articulated ideas on vocal pedagogy. For the purposes of this study, we will address a specific aspect of voice training that Gandharva discusses in the said interview, namely his use of 'closed pronunciation' as a pedagogical tool.

A Brief Survey of Kumar Gandharva's Timbral Aesthetics

Before diving into Gandharva's vocal pedagogy though, it would be fruitful to make some observations about his overall timbral preferences while in performance. Commentators have described Gandharva's voice as 'pointedly tuneful, thin, and quick-moving' [13] and descriptions of his music tend to give a lot of importance to his exceptional command over intonation. His resonance strategies have not been studied objectively, however, and it is hoped that the present study should make some headway in this direction. It is this author's contention that Gandharva's overall timbral goal was to create a sound that contained a consistently prominent 'dark' and 'deep' timbre in <400 Hz range and to balance it with a 'bright' timbre in 2–4 kHz range. The present study focuses on the dark aspects of Gandharva's timbre. Addressing the bright aspect of it is beyond the scope of the present paper since it primarily deals with Gandharva's pedagogical strategies, however, this author's doctoral dissertation addresses this in detail [14].

The following spectrograph (Fig. 1) of the first 40 s of Gandharva's studio recording of Raga Shree amply demonstrates these characteristics [15].

This excerpt also demonstrates Gandharva's preference for the nasal consonant /n/ and the vowels /i/ and /y/, all of which have low first formants and are referred to in phonetics as 'close' vowels as described in [4]. This is an example of Gandharva's preference for low F1 vocalization that results in emphasized lower harmonics. This preference is visible across most of his available recorded performances, such as [15], which has ample instances of the same. Given that the above excerpt is from

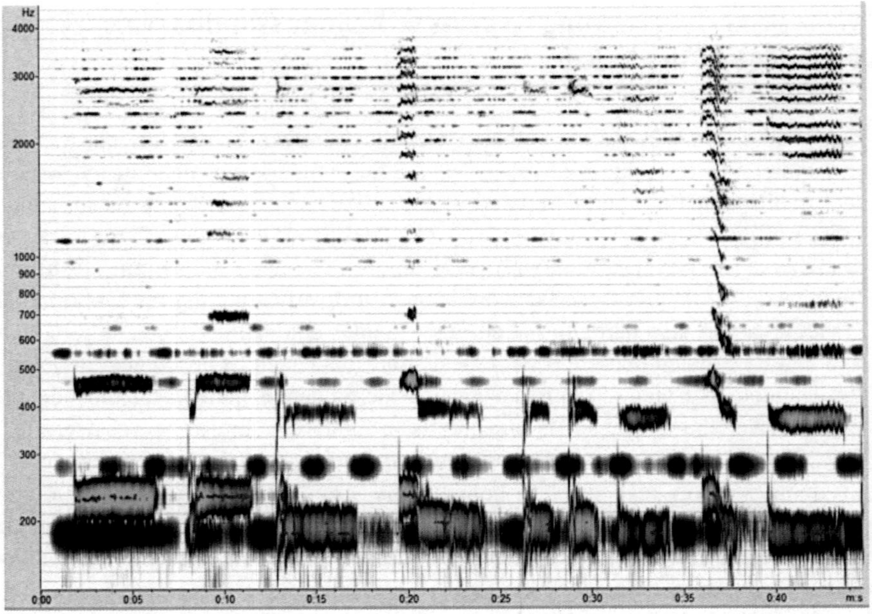

Fig. 1 A spectrograph of an excerpt from Kumar Gandharva's recording of Raga Shree

his initial *ālāp* of the *rāg*, it is reasonable to postulate that one of the goals of his vocalization is to achieve his desired timbre before he launches into full performance. It must be noted that experimenting with timbre and resonance was itself an important performance strategy for Gandharva; therefore, it is not uncommon to find this kind of close vocalization throughout Gandharva's performances, not only in the initial *ālāp*.

The following discussion will attempt to correlate this strategy with the voice training methods he taught to his disciples and articulated in interviews.

Mukhabandī—Closed Pronunciation as a Pedagogical Tool

Introduction

The following is a translated and paraphrased excerpt from Gandharva's interview in [11, 12].

In this interview, Gandharva makes the claim that Indian vocalists are unable to eliminate nasality from their singing voices. While vocalists keep talking about the 'pure' *ākār*, the sung /ɑ/ vowel that is considered fundamental to Hindustani vocalization, Gandharva claims not many of them are really able to produce it without it becoming nasal. While Gandharva says that nasality is not necessarily a bad thing, vocalists ought to practice it separately to master it, so that it doesn't interfere with their singing where it shouldn't. Gandharva then goes on to describe the idea of a resonant voice or a voice that has *nād*. He says that in order to achieve this, vocalists must practice and master singing with closed vocalization, which he refers to using the terms '*banda uchchār*' or '*mukhabandī*'. These terms literally mean singing with a closed mouth. In a particularly important statement, Gandharva notes that while vocalists are able to sing the *mukhabandī* in some places, they are not usually able to 'make this sound scream' [12]. He then goes on to claim that a pure *ākār* will be produced only when such mastery over closed pronunciation is achieved.

In order to train for a resonant voice, and to achieve a 'pure' *ākār*, Gandharva then urges vocalists to 'truly love' the ā, n, m, and the ŋ (voiced velar nasal consonant) sounds, and insists that it is in such closed vocalizations that the resonance of the consonants and vowels resides. He, therefore, advises vocalists to practice *mukhabandī* throughout their singing range. As an analogy, Gandharva likens closed phonation to striking a traditional brass bell, in which the resonation of the resulting sound is not under one's control after the bell has been struck. It has its own natural quality. Gandharva uses this analogy to suggest that the beginning, middle, and end phases of vocalization are important and that most vocalists tend to begin in the middle and think they are at the beginning [11, 12].

From the perspective of vocal acoustics, the above excerpt, though occasionally cryptic, is laden with many ideas that our interdisciplinary approach can fruitfully unpack. We must first remember that when talking critically about nasality, Gandharva refers to vocalists of his generation or of the generation before him. Discussing the alleged nasality of other vocalists is beyond the scope of this study, but it is clear that Gandharva is attempting here to talk about two kinds of nasality, one that is undesirable (as an affectation in singing) and one that is desirable (as a pedagogical strategy). Insights from within the domain of vocal pedagogy can provide some clarity.

Ingo Titze separates the phenomenon of nasality into what he calls 'nasal murmur' and 'nasal twang'. He defines nasal murmur as a sound that has predominantly low-frequency content (200–300 Hz) that is caused by the propagation of acoustic energy into and through the nose. Titze also specifies that nasal murmur is the sound of the nasal consonants /m/, /n/, and /ŋ/, which are the same consonants Gandharva discusses and demonstrates in the excerpt above. On the other hand, Titze defines nasal twang as being 'an acoustic resonation in the laryngeal vestibule', the resonance of which is high-frequency dominant (2500–3500 Hz) and is radiated mainly from the mouth [16].

Clearly then, Gandharva's advocacy of practicing nasal consonants is an advocacy of practicing nasal murmur. This is a common resonance training strategy in Western vocal pedagogy and has been documented as providing two important benefits: to habituate the singer to the sensory perception of a resonant voice and to give rise to an inertive vocal tract (IVT). An IVT is a vocal tract configuration in which the tract is occluded or blocked in some way such as by closing the mouth, giving rise to inertive reactance which assists the vocal folds in their vibration. This leads to efficient vibration that allows the singer to vocalize with minimum effort [16].

Although the acoustical understanding of these training strategies is a recent phenomenon, the strategies themselves are old and well established in the various European schools of singing. While it is unlikely that Gandharva was directly aware of these Western approaches to voice pedagogy, the possibility cannot be completely ruled out, given that his teacher and chief mentor B. R. Deodhar is known to have traveled to the West and exposed Gandharva to Western artistic thought. Deodhar was also a known expert on vocal pedagogy and the author of a well-known book on voice training, but though his study of Western vocal pedagogy began before Gandharva came to him, it really matured only after Gandharva finished his apprenticeship with him and relocated to Dewas in Madhya Pradesh. Nonetheless, it is not surprising that a singer of the caliber of Gandharva should have discovered and adopted these vocal strategies through his own experimentation, especially given his physical disability and his attempts to overcome it by developing his vocal technique.

Gandharva discusses effortless singing elsewhere in the same interview, but this is beyond the scope of the present paper; although some comments will be made on this aspect below. Our immediate concern is the acoustic character of the *mukhabandī* sound, its effect on the open-mouthed singing voice, and what it can tell us about Gandharva's acoustic strategy. Fortunately, we have, in the interview cited above, Gandharva himself demonstrating this technique.

Analyzing and Interpreting Kumar Gandharva's Timbral Strategies

As is visible in the spectrograph below (Fig. 2). Gandharva produces the nasal consonant /m/ from 8:00 to 8:01 min and then transitions to an (approximately) /ɜ/ vowel. Beginning a vocalization using *mukhabandī* and transitioning from it into an open-mouthed one is a pedagogical strategy that Gandharva taught his students and encouraged them to practice [17]. This closed-to-open vocalization also features prominently in Gandharva's concert performances as seen in the Raga Shree excerpt above. It is thus safe to say that this strategy was important to him and analyzing the acoustic connotations of it would be fruitful.

Spectral analysis allows us to make a number of important observations about the harmonic content of this clip, and to attempt to draw some conclusions thereby:

- H1 is quite prominent in the *mukhabandī* area, and only loses a little intensity in the /ɜ/ area.
- H2 remains consistent throughout the clip.
- H3 is deemphasized in the *mukhabandī* area and gains prominence in the /ɜ/ area.
- H4 and H5 remain largely consistent, though H5 loses some energy in the /ɜ/ area.
- Most of the visible harmonics above H5 are subdued in the *mukhabandī* area and only gain prominence in the /ɜ/ area.
- In the /ɜ/ area, harmonics in the 2–4 kHz region receive a boost. A similar boost is also more explicitly visible in Fig. 1 above. This is the area identified with a nasal twang.

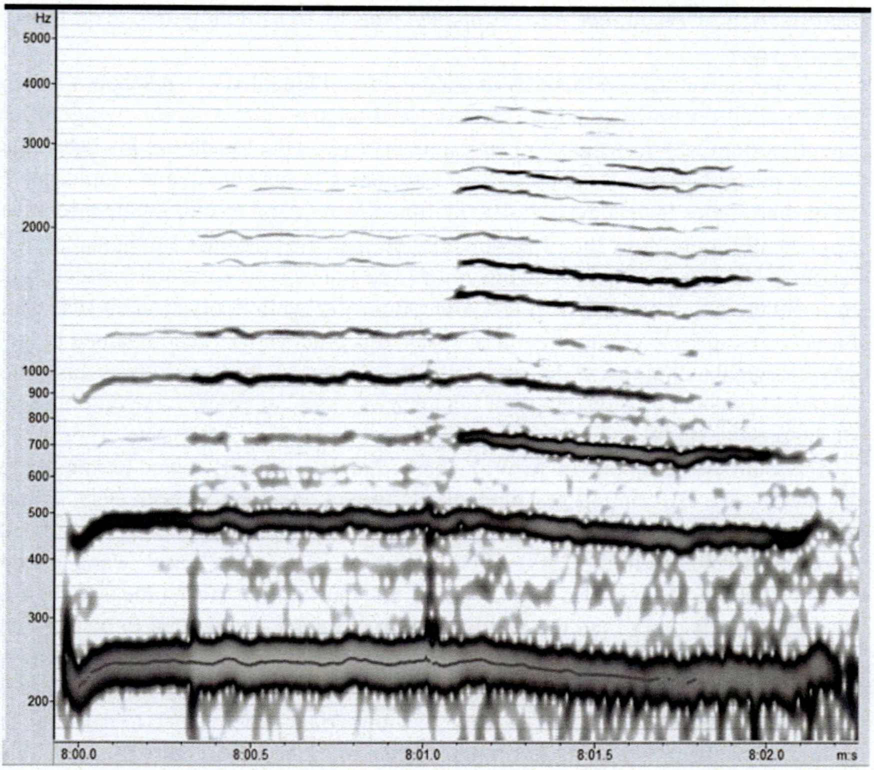

Fig. 2 Spectrograph of Kumar Gandharva demonstrating a transition from an /m/ nasal consonant (closed-mouth vocalization) to an (approximately) /ɜ/ vowel (open-mouth vocalization); *Source* https://youtu.be/UvXOWN7Duc0?t=478 (at 7:59–8:02 min) [12]

We can see here that the /m/ consonant has caused the first formant (F1) to drop to below 300 Hz which is consistent with the observations made in [16], discussed above. This is apparent in the above spectrograph, given the intensity of H1 (which is at 242 Hz). Apart from boosting H1, an occluded vocal tract with a low F1 also substantially de-emphasizes upper harmonics, since acoustic energy gets concentrated in the lower parts of the spectrum. We know from [7] that harmonics above H5 contribute roughness to the overall timbral construct, while those above H9 do not resolve into the fundamental, thus furthering this perception of roughness (Table 1). As seen in the above sample, the *mukhabandī* sound severely de-emphasizes harmonics above H2, generating, as discussed in [7], a relatively 'pure' tone that lacks both roughness and unresolved harmonic content. When Gandharva later opens into the /ɜ/ sound, we find that although harmonics H6, H7, and H10 onwards have received emphasis which is necessary to form the intended vowel, they are still relatively subdued, while H1 and H2 continue to be prominent.

This leads us to the important conclusion that Gandharva uses *mukhabandī* in order to train his voice to emphasize his lower harmonics, **and then aims to carry**

that particular resonance into his open vowels by transitioning into them *from* his *mukhabandī*.

We can now correlate this new understanding of Gandharva's resonance strategy with his well-documented mastery over intonation to postulate the following: At least one important way in which Kumar Gandharva was able to achieve his tunefulness was to deemphasize a large number of harmonics in his signal, and emphasize the lower harmonics instead, especially the fundamental, and to thus create a relatively acoustically 'pure' tone that was well aligned with the reference fundamental of the tānpurā. He was thus able to avoid rough and unresolved harmonics which might have otherwise interfered with the perception of a well-resolved pitch that Gandharva was known to create. Needless to say, this practice was accompanied by good physiological technique, which allowed him to *maintain* this tunefulness throughout his vocal range. The physiology of Gandharva's singing voice is beyond the scope of the present study.

It would be pertinent to mention here that emphasized lower harmonics are an incomplete description of Gandharva's timbre. As seen in both Figs. 1 and 2 above, Gandharva also creates a characteristic 'ring' in his voice by significantly boosting harmonics in the 2500–3500 Hz region. This is the 'nasal twang' component discussed above. Also significant is the fact that at most times, though not always, regions other than the nasal murmur and nasal twang tend to be *deemphasized* in Gandharva's voice. Thus, the resultant timbral construct is one that has two distinct components: a clearly identifiable 'dark' tone, balanced by a clearly identifiable and distinct 'bright' tone. As Table 2 shows, dark timbres tend to have an /u, o, ɔ/-quality to them, while bright timbres have a more /æ, e, i/ quality to them. While this needs further investigation, it could be hypothesized that Gandharva's universal employment of and frequent dependence upon his characteristic 'ye' sound [13] was an attempt to create just such a balanced timbre. While it is beyond the scope of this paper to verify and substantiate this claim, the present author's many years of experience studying the music of Gandharva suggests that this is indeed an acceptable description of Gandharva's tonal signature. It may be mentioned in passing that this balanced dark-bright timbre is starkly reminiscent of the concept of chiaroscuro resonance, a well-established vocal ideal in Western Operatic singing [4]. A more detailed investigation of Gandharva's timbre that takes this unexpected parallel into account is forthcoming.

The account of Gandharva's timbre we have thus arrived at also allows us to understand commentators' descriptions of it in a more concrete fashion. We can now say that the traditionally used terms 'pointed' and 'thin' describe Gandharva's emphasis on particular bands of closely spaced harmonics that are especially boosted at the cost of others, in contrast with more 'spread out', 'broad' voices such as that of Ustad Faiyaz Khan that gives equal emphasis to a larger number of harmonics. The example of Faiyaz Khan is used here only to establish contrast. Many recordings of both musicians are available online, and a simple aural comparison will make this point amply clear.

Using Computational Methods to Problematize Information in the Oral Tradition: Some Inferences and Observations

The above insights are important because they allow us to understand the reasons behind Gandharva's pedagogical methods in concrete, measurable terms. At the same time, acoustic analysis also allows us to problematize Gandharva's ideas in important ways. A case in point is Gandharva's claim in the quote above that this sort of training will allow the student to produce a 'pure' *ākār*. The conventional understanding of the 'purity' Gandharva mentions would be phonetic—a 'pure' *ākār* would mean a vowel sound that is unambiguously an /ɑ/, such that it could not be confused with any other vowel. However, commentators have often pointed out that a phonetically pure /ɑ/ was conspicuously *absent* in Gandharva's singing [13]. We know that the /ɑ/ vowel has a much higher first formant (around 1000 Hz) than the nasal consonants, and using it might go against Gandharva's timbral goals to some extent. This is a possible explanation for his avoidance of it in performance.

It is this author's contention then that **Gandharva's use of the term 'pure *ākār*' connotes an *acoustic* rather than a phonetic purity**—the kind of purity present in a timbral construct that lacks roughness, in the sense embodied by Howell's use of the term.

As a disclaimer, it must be mentioned here that within the context of Khayāl music, the usual reference for a true, pure *ākār* is the vocalization of Jaipur Gharana vocalists such as Kesarbai Kerkar. It is in comparison with *their ākār* that Gandharva or others like Abdul Kareem Khan are described as *not* having a pure *ākār*. However, listeners accustomed to Gandharva's vocalization are likely to find Gandharva's *ākār* entirely satisfactory.

With reference to Gandharva's comments on desirable and undesirable nasality, we can now say that while he is not an advocate of a nasal affectation in the singing voice, he is certainly an advocate of using nasal sounds, particularly Titze's 'nasal murmur', *as a tool* to achieve his particular timbral goals—those of a close, lower-harmonics-dominant 'pure' timbre.

With regard to Gandharva's analogies of a struck bell and a tossed pebble, these can be correlated with our understanding of how acoustic pressure works in an occluded or semi-occluded vocal tract, or in other words, of the non-linear source-filter model of voice production mentioned above. We have already seen in brief above how an IVT assists vocal fold vibration and encourages effortless singing. I contend then that this is what Gandharva means when he says that 'once you start singing, you don't have to control the resultant sound'. What he means is that the produced sound will *resonate without extra muscular effort*—a common phenomenon in IVT-based vocalization. Additionally, it is also possible to infer that Gandharva's comments about the importance of the 'beginning, middle, and end' address the vocalist's glottal attack. His comment that we 'begin in the middle and think we are at the beginning' [12] creates the image of a vocalist trying to 'control' his timbre and achieve volume during vocalization through pressed phonation, a mode where the vocal folds are tightly adducted, resulting in a harmonically rich timbre, and also more effortful singing caused by increased subglottal pressure [4], rather than creating an IVT configuration and leveraging the resulting resonance to achieve those goals. This is admittedly a hypothesis and would need further investigation.

This would also explain Gandharva's advocacy of closed vocalization as a tool to train for resonant vocalization—he clearly uses the term resonance in its literal, acoustic sense to denote that certain harmonics are *reinforced* by particular vocal tract configurations, instead of using it in the general, vague sense it is conventionally used among vocalists. Deliberately creating boosts in intensity by encouraging particular harmonic-formant interactions is a common way of achieving volume without expending too much effort, a phenomenon often referred to as 'volume for free'. This is what Gandharva means when he says he uses *mukhabandī* to make the sound 'scream' or to 'howl like the wind' as his disciple Shubha Mudgal puts it [18].

Conclusions and Further Research

We have seen how we can use an interdisciplinary approach that combines knowledge from the disciplines of musicology, vocal acoustics, psychoacoustics, and voice science or vocal pedagogy together with computational tools to develop a nuanced understanding of the information available in the oral tradition. While the present study focuses on a Hindustani classical musician, it is pertinent to point out here that unlike *rāg*-identity, musical timbre is *not* a traditionally theorized construct and is thus applicable to all genres of music. This opens up the research opportunity of applying computational timbral analysis even to the genres of Indian music that do not have classical status.

Timbral analysis is a particularly potent approach that allows us to demystify a lot of the informal, oral discourse that surrounds voice use in traditional Indian music. While the present study has attempted to explicate the resonance strategy of a particular master musician, it in no way makes the claim that this is the only, or the model resonance strategy Hindustani musicians employ. It is, on the contrary, the case that this genre of music, like many other genres, encompasses tremendous diversity in the timbral choices and aesthetic vocal goals its practitioners employ.

Timbral analysis of this kind opens up a number of intriguing possibilities for further research, some of which have been outlined below:

- Clearly understanding and defining the diverse vocal goals of diverse musicians and linking them to their larger musical aesthetic
- Tracing timbral lineages by identifying similarities in vocal timbres
- Identifying particular physiological and acoustic configurations that are markers of particular aesthetic/stylistic idioms
- Examining acoustic/physiological reasons behind resonance strategies that have traditionally been considered important, such as the *ākār*
- Correlating musicians' timbral goals with their structural choices made in the course of performance

- Working with contemporary performers/teachers/students to document their resonance strategies

It is hoped that the present paper encourages further work in these directions.

Acknowledgements The author would like to thank Dr. Ambuja Salgaonkar, Head, Dept. of Computer Science, University of Mumbai for her keen interest in this work and her useful feedback. Thanks are also due to Pt. Satyasheel Deshpande for his musicological insight and expert comments, and Dr. Nikhil Govind, Head, Manipal Centre for Humanities for his unwavering support.

References

1. H. Powers, Classical music cultural roots, and colonial rule: an indic musicologist looks at the muslim world. Asian Music **12**(1), 5–39 (1980). https://doi.org/10.2307/833796
2. K. Schofield, Reviving the golden age again:'classicization', Hindustani music, and the Mughals. Ethnomusicology **54**(3), 484 (2015). https://doi.org/10.5406/ethnomusicology.54.3.0484
3. S. Rao, P. Rao, An overview of Hindustani music in the context of computational musicology. J. New Music Res. **43**(1), 24–33 (2014). https://doi.org/10.1080/09298215.2013.831109
4. K. Bozeman, *Practical Vocal Acoustics: Pedagogic Applications for Teachers and Singers* (Pendragon Press, Vox Musicae Series, 2013)
5. https://www.sygyt.com/en/
6. I. Howell, *Parsing the spectral envelope: Toward a general theory of vocal tone color* (New England Conservatory of Music, 2016)
7. I. Howell, Necessary roughness in the voice pedagogy classroom: the special psychoacoustics of the singing voice. VOICEPrints J. N. Y. Sing. Teach. Assoc. **14**(5), 4–7 (2017)
8. https://www.youtube.com/watch?v=Ud3yHDiP42o
9. V. Deshpande, *Ālāpinī*, 1st edn. (Mouj, 1979)
10. V. Deshpande, *Between Two Tanpuras* (Popular, 1989)
11. K. Gandharva, M. Bhatavdekar, *Kumar Gandharva: Mukkam Vashi,* 2nd edn. (Mauj, 2007)
12. K. Gandharva et al., *Mukkam Vashi*, November 1992; Audio Recording, Archives of the Manipal-Samvaad Centre for Indian Music, MAHE Manipal.
13. A. Ranade, *Some Hindustani Musicians: They Lit the Way!* (Promilla & Co., Publishers in association with Bibliophile South Asia, 2011)
14. S. Deshpande, The alterity of Kumar Gandharva: examining musical otherness in the tradition of the Hindustani Khayal (thesis). Manipal Acad. High. Educ. (2022). http://hdl.handle.net/10603/447675
15. K. Gandharva, *Sangeet Sartaj Vol. 1: Raga Shree*; Music Today CD-A02057/58 [CD, Compilation, 2002]. https://www.youtube.com/watch?v=AnsBK4SLvHg
16. I. Titze, Acoustic interpretation of resonant voice. J. Voice **15**(4), 519–528 (2001). https://doi.org/10.1016/S0892-1997(01)00052-2
17. Personal interview with Satyasheel Deshpande, disciple of Gandharva, 25/2/2021
18. S. Mudgal, *The Journey from Voice to Gayaki, Kaljayee Kumar Gandharva*, ed. by R. Inamdar-Sane, K. Komkali, 215-21 (Rajhans Prakashan Pvt. Ltd. 2014)

Software Assisted Analysis of Music: An Approach to Understanding Rāga-s

Sandeep Bagchee

Introduction

Music is an art form that manipulates tonality in sound, so as to convey an aesthetic message that moves us. Appreciating a piece of music and understanding it is a complex process that is largely subjective and depends on the musical conditioning and exposure of the listener. This process is further impeded by the transient and ephemeral nature of a live performance, which then also makes it dependent on memory. Analysis of music and its comprehension, thus poses problems quite different from those of other arts.

Technology has been able to reduce these constraints to some extent. The replicability of music brought about by the ability to record music and reproduce it, has reduced complete dependence on memory; the availability of computers and their use for analyzing music has reduced the subjective element in this process. The use of computers and computational methods to analyze and understand musical structures has given rise to the interdisciplinary area of computational musicology. With further advancements in hardware, the availability of computers with greater processing capability and more sophisticated software, this technique has evolved greatly since its inception [1, 2]. In India, too, computational musicological studies of Indian musical forms have been carried out, albeit somewhat less as compared to the work done on genres of western music [3].

As *rāga* is the fundamental concept underlying the music of the Indian subcontinent, it is not surprising that issues related to it have been the focus of computational music studies, covering topics such as automatic *rāga* recognition, extraction of melodic patterns using data mining, detection and classification of melodic motifs etc. [4–11]. These explorations focus primarily on the techniques, such as hidden Markov

S. Bagchee (✉)
Pune, India
e-mail: sandeep.bagchee@gmail.com

A. Salgaonkar and M. Velankar (eds.), *Computer Assisted Music and Dramatics*,
Advances in Intelligent Systems and Computing 1444,
https://doi.org/10.1007/978-981-99-0887-5_3

models and data mining, for extracting musical features from recordings. However, the focus of computational music studies has not, so far, been on advancing musical understanding per se. There has consequently not been adequate interest in using computational musicology to investigate relationships between prevailing musical knowledge and contemporary performance practice [12]. Despite highlighting the need for such a direction in computational musicology, its current focus continues to be primarily on statistical analysis as is evidenced by a recent study in Hindustani music [13].

It is our view that computers and the available software for musical analysis can greatly aid in promoting a general understanding of raga, particularly how this basic musical form is perceived and comprehended by the lay listener, and therefore, it is this area which should be the real focus of computational musicology in India. This paper is a modest effort in this direction. Its examination is limited to aspects of how the raga is understood by the 'ordinary' listener, one with some exposure to the musical idiom of raga music but not necessarily an 'experienced' listener. This categorization into ordinary versus experienced listeners is adopted here for convenience. But it seems a valid one, since existing literature has already categorized listeners as experienced, ordinary, uninitiated or even 'illiterate,' based on their knowledge of or degree of exposure to music [14, 15].

The Rāga

It is necessary, at the outset, to outline our understanding of the *rāga*, the focus of our enquiry. Broadly speaking, a *rāga* provides a tonal framework for composition and improvisation [16]. Each *rāga* contains certain specific tones from the scale and the order in which they are to be sung or sounded is fixed and broadly specified by ascent and descent patterns. In addition to their order of usage, the manner in which these tones are to be sung including their duration and extent of usage, the degree of emphasis to be given, the intonation and 'attack', the ornamentation to be used are all specified. This results in a hierarchy among the tones comprising the *rāga*. Further, as a consequence *rāga*-s usually have specific tonal combinations or phrases which are often regarded as unique to them and termed as key phrases or motifs (*pakad*), as well as more extensive melodic tone patterns (*chalan*). Even though there is no single canonical version for any one *rāga* and different interpretations exist with respect to most *rāga*s, there is a broad area of convergence on what constitutes the salient features of any particular one. Each *rāga* has certain distinguishing characteristics that constitute its identity. All these features are usually employed to identify and distinguish between *rāga*s, particularly when they employ the same tones. These features constitute a structural view of the *rāga* as a form of melodic organization.

In addition to being a melodic structure or form, the *rāga* is also a cognitive category; a 're-identifiable particular' that we recognize by a sense of the "same again", in which we place all performances and instantiation of melodies that appear to us as being similar, in a common category. We therefore need to also understand how the *rāga*, in general, is cognitively recognized so that it functions as a special kind of "'re-identifiable particular' in the world of sound" [17].

The concept of schema, the building block of cognition, which refers to a cognitive framework that helps organize and interpret information, is a useful tool in this pursuit. In general, our attention is guided by schemata which are the knowledge of structures developed in our experience of the world [18]. In the case of music, listeners intuitively analyze it through its features (such as tones, intonation, ornaments) using them as cues in the selection of schemata, and schemata, in turn, serve as guides in the detection of further features in what is, basically, an iterative process. When a feature is presented, we attempt to find a context to it by regarding it as a partial instantiation of one of several possible schemata and as more features are perceived, alternative schemata can be eliminated and the most likely one selected. Schema theory asserts, consequently, that once distinctive features of a schema are instantiated, we actively seek out the remaining features [19].

Musical schemata can be of different sorts. Some may incorporate general knowledge of properties or features common to many pieces of music, e.g. our knowledge of the *sargam*, the diatonic scale underlying Indian music, or the schema of a *genre* like *khayal* or a style particular to a *gharanā*. There can also be even more specific schemata which encapsulate knowledge of tonal relationships within a particular melody, such as a melodic contour. Thus, there can be several different types of schemata that the mind refers to in the process of listening to a piece of music—including melodic and rhythmic ones. In the perspective adopted here, we propose viewing the *rāga* as a melody schema, as this would help us understand the process of cognition by which a specific *rāga* is recognized on hearing a melodic piece and then categorizing it.

Although *rāga* music involves melody as well as rhythm, our focus will be, principally, on melody. We will begin first by examining how melody is perceived and comprehended and based on this, make some suggestions as to what constitutes a melodic schema and how the melodic schema functions in the case of the *rāga*. (The element of rhythm could conceivably also be built in subsequently, but for the present, we will focus only on melody in Hindustani music).

Melodic Perception and the Rāga

Melody is the most ubiquitous form of musical structure and is formed, in essence, by a connected series of tones of various pitch-values, duration and rhythm. It arises because of our ability to hear a sequence of pitches coming together with a distinctive shape. This ability is a manifestation of the brain's quest for coherence in the stimuli it receives. Further, underlying the listeners' perception of melody are tonality and contour, two factors that play a principal role in the cognitive processing of music, particularly how we remember it [20]. Tonality refers to the hierarchical organization of tones around a single reference pitch, the tonic or anchor tone. The contour of a melody is the shape that arises as melody moves from tone to tone: rising if the subsequent tone is higher in pitch, or descending if it is lower, or leveling out if the tone is held for some time. Indian art music is tonal with the *Śadja* as the reference tone. Further, much of the musical experience in Indian art music is articulated through terms such as ascent and descent, crooked or zig-zag or meandering movements. These are terms which refer to melodic contour.

The contour of a melody, how it rises and falls in pitch provides important clues for cognition and recognition. In the case of tonal music, each tone is perceived as being related to the other tones. At its most basic, the relationship between the preceding and succeeding tones of a sequence depends on their relative positions in the scale and these tone-to-tone relations of a rise or a fall come together to provide the overall shape or contour to the tone sequence, and this appears to be central to our experience of melody. Contour remains the basic way through which the melodic pattern brings itself to our attention. Alongside this, there is a deeper structure to the melody that arises out of the intervallic relationships, that is the distance between each of the tones and the tonal center, and the mind works out this pattern of hierarchies, albeit subconsciously [21, 22].

Explanatory contour models in cognitive music theory are divided into two categories, local models which focus on individual interval content of contours and the global models which explain phenomena such as 'gap-filling' where large tonal jumps are followed by a reversal of melodic direction and the characteristic melodic arch found among the general patterns of melodic contour. Studies suggest that melodic contour can be reliably characterized by both individual relations within the contour as well as by more global contour parameters with the former approach applicable to short melodic contours while the latter providing better explanations when applied to longer melodies [20]. Short term melody recognition is strongly influenced by contour information but becomes less precise in memory as melody length increases. It is, thus, clear that although contour is crucial for melodic memory it functions in conjunction with other important melodic features such as tonality.

We have suggested above that viewing a *rāga* as a melodic schema may provide insights as to how it is recognized and understood. However, a melodic schema must go beyond and map something more abstract than merely specifying pitches and time intervals. Consequently, it has been proposed that a melodic schema comprises the contour plus the specification of the tonic and the tonal scale. Alternatively, it has been suggested that schemas of familiar tunes are mapped as relative pitch chromas rhythmically organized which can be accessed by labels and melodic features such as contour [21]. Thus, analyzing the contours of performances of specific *rāga*-s could provide some clues to their functioning as schemata.

This paper will, therefore, focus principally on the melodic form and contours of selected *rāga*-s using pitch-extraction software tools that allow the melodic contours to be traced. For this purpose, some well-known commercial recordings of these *rāga*-s will be used to trace their melodic contours. These will then be examined to try and identify the common features among different performances of the same *rāga* in an attempt to understand the extent to which these conform to the prescriptive requirements, as also the extent of deviation from them. In addition, an attempt will be made to identify the features common between different performances of the same *rāga,* in an attempt to understand how an ordinary listener is able to place these different instantiations in a single category and thereby grasp the concept of the cognitive schema underlying that particular *rāga.*

We next outline how a raga's contours are derived from various recordings of its performance, so as to identify the tones that are sung and their duration, as well as the intonation and modifications rendered through the use of various musical ornaments and embellishments. The contours also tell us how the melody moves from tone to tone.

Graphical Depiction of Melodic Contour

Attempts to mechanically transcribe music in a graphical form, the melograph, where the pitch is plotted against time, date back to the beginning of the twentieth century but gained currency in the mid-fifties with the Seeger melograph [23]. In the case of Indian music, the issue of pitch perception and pitch extraction in melodic music was examined and a Fundamental Pitch Extractor (FPE) was built and coupled to a computer system thus giving rise to the Melodic Music Analyser (MMA) in the nineteen eighties, enabling one to picture and measure melodic lines with reasonable accuracy [24]. Subsequently, the system was refined with the availability of greater computing power and more advanced soft-ware, including the availability of PRAAT, a software for phonetic research which can be used for making graphs. These developments in graphical representation led to the Automated Transcription of Indian Music (AUTRIM) project where moving melographs show the melodic line or contour in a graphical form and provide us with a more objective and more importantly, a visual means of analyzing *rāga* performances which have been specially recorded for the project [25]. However, although PRAAT can be used for musical analysis, it provides more reliable results with separate recordings, that is, with a dedicated microphone for voice and a separate one for the accompaniment [26].

Instead of PRAAT, this study used the Melodia plug-in, which automatically estimates the pitch of a song's main melody, that is, the fundamental frequency corresponding to the pitch of the predominant melodic line of a piece of polyphonic music [27]. The Melodia vamp plug-in was used directly in Python, using a code written for tonic normalization and transcription to tone sequence (*sargam* scale). The extracted pitch values of the main melody, along with the sound file of the recording, were imported into Sonic Visualizer, an application for viewing and analyzing the contents of music audio files [28]. This facilitated an understanding of specific raga performances, since it allowed access to the raw pitch curve, and enabled observing the nuances of pitch movement.

The Rāga-S

In this paper two sets of *rāga*-s have been selected for examination. While the *rāga*-s in each set share the same tonal material, the authoritative sources outlining the *rāga*-s stress that care should be taken while performing them, so that the identity of each *raga* is clearly established. There should be no confusion in the listener's mind in evoking the clearly separate images, of one from the other, sought to be created when rendering them. The current understanding of each of these *rāga* -s in terms of their tonal material, their ascent and descent, their motifs (*pakad*) and characteristic movements or tonal combinations (*chalan*) is first outlined, relying on the theoretical material, namely, the codification of these *rāga* -s in the writings of various sources and authoritative writers. The identities of these *rāga*-s as laid down in these works, the 'prescriptive', that is a 'blueprint of how a specific piece of music shall be made to sound' is then followed by an analysis of actual performance practice, the 'descriptive', 'a report of how a specific performance of any music did actually sound' [23].

The two groups are (1) Mārvā, Pūriyā, and Sohanī, and (2) Bhūpāli and Deśkār. In respect to both sets of *rāga* -s, we examine if there is a possibility of their images being confused with each other and what features differentiate the raga-s within each set. Secondly, we can examine if based on these identifying features, we can establish the schema of each *rāga.* The performances chosen for analysis are commercial recordings of these *rāga*-s, some of which have been transcribed by Magriel, and indeed, have been selected on this basis, because his transcription of the *bandiś*-es provides a reference point to further validate our analysis [29].

It should be mentioned, in this context, that there are some difficulties in interpreting these melodic contours because the time–values of tones are not quantified, however approximately, for any rāga. Thus, while tones are referred to as standing tones, *nyāsa* or rest tones etc., it becomes difficult to decide as to whether the tone is a standing tone or a grace note. Inevitably, the importance of tones has to be determined on a relative basis and subjective judgment exercised, based on auditory perception rather than on visible and empirically verifiable characteristics such as the time for which the tone is held. Consequently, this may not result in consistent outcomes. Therefore, besides relying on the available transcriptions, we have also relied on interpretations from earlier studies using melograms/melographs as regards features such as intonations, standing notes, as well as ornaments such as *mīṇḍ*-s, *gamak*-s etc [30–33].

Mārvā, Pūriyā and Sohanī

The Prescriptive

These *rāga*-s share common tonal material, namely the six tones *Sa*, *re*, *Ga*, *Ma*, *Dha*, *Ni*; the fifth tone *Pa* being prohibited. Most sources caution the performer to clearly maintain the distinction between the Mārvā and Pūriyā and clearly establish the identity of the *rāga* being performed. In the case of Sohanī, there is less chance of confusion, since it is quite distinct in having a different melodic character.

Even though there is agreement about the constituent tones, the sources differ about the order of the tones in ascent and descent, though these do not by themselves characterize the melodic movements [34]. The differences arise not only because of the differences in the emphasis on some of the tones but also because of the context of the specific tone combinations in which these are used, as well as the distinct melodic phrases that arise as a result. We need to see if these are present in practice.

The initial ascending phrase, *Ṇ r G*, is similar for both Mārvā and Pūriyā, except that the tone *re* is required to be emphasized in Mārvā in contrast to the emphasis on the tone *Ga* in Pūriyā, without pausing on *re*, thus, differentiating the two ascents. On the other hand, Sohanī commences with *Ṇ S G M D* to reach the upper tetrachord. Although both Sohanī and Pūriyā, contain the combination *G M D N*, in the former the *Ṇ S G* precedes this common note combination while the latter has *Ṇ r* prefixed instead. Mārvā is often regarded as the parent *raga,* with three definitive tone combinations, namely, *Ṇ r_ G r_*, *D M G r G M D*, and *M D N D N r_ N_ D_*. The other two *rāga* -s, namely Sohanī and Pūriyā can be derived from these tonal combinations [34].

In Mārvā, the emphasis is on tones *re* and *Dha*, *Ni* is a weak tone while *Sa* is skipped both in the ascending and descending sections. Even the high *Ṡa* is not approached directly from *Ni* but obliquely by descending to *Dha* before touching *Ṡa* [35]. Other commentators mention that the resolution on *Sa* is to be avoided or rather delayed to the extent possible as this delay provides the *rāga* with its characteristic mood of tension. However, the tone cannot be totally omitted as this would result in *re* being heard as the ground tone and, thus, convey the impression of a different *rāga*, Mālkauns.

Thus, there is an overwhelming dominance of the tones *re* and *Dha* in Mārvā, and these have to be stressed in the descending phrase *D_ M \ r'* (*mīṇḍ*) to distinguish it from Pūriyā which has the phrase $D\ {}^{G}M\_\backslash G$ [36]. Some sources state that ornaments are not used in Mārvā while they appear in Pūriyā.

Pūriyā is said to consist of many more tone combinations compared to Mārvā. The tones *Ga* and *Ni* are strong in Pūriyā; the dominance of *Ga* results in both *re* and *Dha* being relegated in importance. The phrases $G\ M\ D\ N\_\ N\ {}^{N}M\_\ G$ and *M D_G M G* are said to constitute the soul of the *rāga* and while *Ni* is strong in Pūriyā, it is often skipped in the ascent to the high *Ṡa* as in Pūriyā. The two features said to be important in Pūriyā are the "swoop from *Ni* to *Ma*" and the *Dha-Ga* conjunction [36]. Others point out that the glides *N/M* and *M\N* are important characteristics for

Pūriyā, along with *M\G* and *N/M\G* [25, 37]. The conjunction has been indicated as the phrase N^{D}*\M_G* in one source while others show it as D^{G}*M_\G* or as the phrases *G M D G M G* and *M D N r N\DM-G* [38]. Thus, from these various interpretations, it can be concluded there is descending *mīṇḍ* whether from *Ni* to *Dha* or *Dha* to *Ma* or *Ma* to *Ga* via the tone *re* which is an "essential" feature of Pūriyā though there appears to be no consensus regarding its exact form.

In Sohanī, on the other hand, the melodic movement is in the middle and upper registers. We have already covered most of the features that distinguish it from Marva and Puriya. Its characteristic *chalan* is shown as: *G M D N Ś_ Ś ŕ Ś ŕ N D_ N D-G M G* [35]. Sohanī has a direct approach to Sa via Ni, which Mārvā and Pūriyā do not. Also, Sohanī has the note combination D G M G, except that in Sohanī this is approached from the tone Ni, that is in descent, while in Pūriyā this is approached from *Ma* below. Further, while the *vadi-samvadi* pair in Pūriyā is *Ga* and *Ni*, it is *Dha* and *Ga* in Sohanī [34].

Analysis of the Performances of Marvā, Pūriyā and Sohanī

Two *bandish*-es in Marvā - "*Piyā morā anat desa gavana kino*" in *Jhumrā tāl* and "*Guru bina gyān na pāve*" in *tīntāl,* sung by Amir Khan, recordings which are largely regarded as exemplar, have been selected. For *Rāga* Pūriyā, the selection consists of three *bandish*-es "*Chhina chhinna bāṭa takata hŭ torī*" in *tīntāl*, by Amir Khan and "*Ghadian ginata jāta*" in *tīntāl*, and "*Pyārī de gara lāgo*" in *vilambit ek tīntāl* by Bhimsen Joshi. For Sohanī, a *bandiś* sung by Malini Rajurkar is included. We will analyze and compare these in order to understand the issues that we have discussed above.

But before commencing this comparison, some observations need to be made. As the transcription or rather the graphical depiction of the melodic contour is done using pitch extraction software and the position of the tones fixed after determining the pitch of the singer's tonic, it is assumed that the identification of the tones is quite precise. However, it is seen that sometimes, the singer does not hold the exact tone, and there is some amount of variation, albeit minor, around the pitch position. In our analysis below, we have chosen to ignore this and if the melodic line is in the vicinity of a tone, we have interpreted it as intoning the tone and not as being "out of tone". This, in fact, is in keeping with our discussion regarding how the pitches are perceived and categorized as tones, despite the prevalent belief that 'good' musicians produce pitch perfect tones. Another problem with interpreting these graphical depictions, is regarding the time value of the tones that appear in the graphs. While the time value of notes poses a problem in Indian art music as these are not precisely specified and only general statements, such as the tone is held or prolonged or rested on are made, this problem gets accentuated when we have a graph with the tone value on one axis and time in seconds on the other axis as we are forced to decide whether a note held for half-a-second or so is to be treated as an emphasized one or not. The entire exercise, thus, becomes relative. Even more difficult is interpreting single points of the graph and deciding whether the notes represented by such points are melodically significant, or whether these should be regarded as grace notes or treated as points of attack or approach to a sustained note from below or from above. Tone values of the time order of 10–15 milliseconds get perceived as tones in the case of pure sound [39]. The high degree of variation in the trace of the graph leads to other problems, for instance, how does one treat a point of inflexion on the graph? With these caveats we proceed to examine these recordings.

Performances of Marvā

Amir Khan (HMV EALP1253/ HMV Cassette SM 95,006 Re-issued as Saregama CDNF150531).

1. *Khayāl (vilambit jhūmrā)*

The performance begins with the tonic being established by being held followed by a drop to *Ḍha* and a rise to *Ṇi* and *re* which is paused upon, thus, singing the phrase *ḌṆr_* which conveys a sense of Mārvā [34]. There is a drop to *Ḍha* from here, a rise to *Ṇi* and a return to *Ḍha*, now held. The subsequent two phrases are ^{D}*ṆS_* where *Sa* is held with touches of the tone above and rising from below and *Ḍ* ^{r}*S*.

The *sthāyī* is taken up with the *mukhṛa* "*Pia more*" sung as *S_*^{r}*Ṇ*$_{Ḍ}$. The first two syllables of *anata* are sung to an extended *Ḍha*, while the last syllable *ta* is sung as an undulating line ascending from below to *Ṇi* to *Ga* to fleetingly touch *Ma*. The next word *desa* is sung in continuation descending to *Ga* from *Ma* and then rising to *Dha* for the first syllable, dropping to *Ma* for the second in a *D/M* glide. For *gavana*, a quick rise from *Ma* to *Dha* is followed by a descent, the syllable being sung as ^{D}*MGr*, the characteristic Marvā phrase, followed by a quick drop to *Ṇi* and *Ḍha* while *na* is sung in an undulating ascent from *Ṇi* to *re*, back to *Ṇi* and *re* before taking *Ga* to sing *kino* as a descent *G\rS* with a long rest on the tonic. This is one of the few instances in which the tonic is highlighted in a descending phrase.

In the second line *nā jānu*, the syllable *nā* is sung in an undulating manner as ascent from *mandra Ḍha* to *Ga* to *re* and a quick drop to *Sa* while *jānu* is sung similarly as an undulating rise from *Sa* to *Dha* with a pause on the last tone before a drop to *Ma*. The two words *kaba ghara* are sung to a *mīṇḍ* from *Ma* via *Ga* to *re* (*M*^{G}*r___*) with *re* extension on the last syllable. This descending phrase although not commencing from *Dha* gives a sense of the *rāga*. For the next word *āve* there is a brief rise from *re* to *Ga* and a descent to *Ṇi* and *Ḍha* before rising to the tonic. This presentation of the entire *sthāyī* is concluded by repetition of the *mukhṛā* followed by the *baṛhat*.

In the first part of the *sthāyī*, the lower octave is emphasized and rising from there to *Sa*, to establish the tonic. The phrase *ḌṆr_* in ascent parallels the middle octave movement by avoiding resolution on the tonic. The frequent usage of *Ḍha* brings out its prominent role.

Development follows largely in *ākār*, initially, in the lower register with focus on *Ṇi* and *Ḍha* but with a few instances of ascent from *Ḍha* to pause on the tonic. The tone *re*, at first fleetingly touched, is later emphasized and forms the upper limit of the development with descents to *Ṇi* and *Ḍha*. Next, the tone *Ga* is touched, followed with descent to *re* giving a sense of the *mīṇḍ* and *Ga* is then taken prominently with prolonged pauses on *re*. *Ma* is next taken in this systematic development and the middle *Dha* follows soon thereafter. While the ascent to *Dha* is stepwise, that is tone by tone without a pause on re, the descent is clearly through a *mīṇḍ D M \r__* and the tone *re* is held without descending to the tonic. This phrase starting and ending with the two main tones also brings out their opposition because of their lack of consonance. Although the tonic is periodically sung in between from time to time, this is in ascent from the lower octave. The tone *re* while taken in ascent, is not paused upon as prescribed.

The next development is through *sargams* with the ascent and descent being very clear, following patterns such as: *ḌṆr G M D M\Gr*, *Ḍ/MG\r __Ṇ Ḍ*, *ṂḌṆr__*, *D M\G r_ G M GrGr ṆḌ*, *ṂḌṆr__ṆḌ*, *ḌṆrGM\Gr_G m G r_*, *GṆr_ṆḌ*, *ṂḌ_Ṇr__*, *ḌMG\r_*, *G\r_ṆḌṆṆrS* interspersed with repetitions of the *mukhṛā* . The focus on the lower octave brings out the gravitas of the *rāga* as well as the dominance of the tones *Dha* and *re*. *Sargams* are followed by *ākār tāns* and variations of the *mukhṛā* as well as other lines sung to *r Ḍ Ṇ G M D P ND_M\Gr_G MDM\Gr* as an ascending and descending arch with *re* being sustained in descent without further descent.

In the *antarā* that follows, the first word *una* is sung basically as *DM_DM_* with a trill at the end rising to *Dha* before descending to *Ma,Ga* and *Re* and rising again to *Ga* and *Ma* the high *Ṡa* reached with the second word *ke* sung to *Dha* and *Ma* with a slight glide. *Darasa* is sung to the *tār Ṡa* extended with touches of *Ni*. *Dekhabe*, begins with the first two syllables sung to this tone before a drop to *Ni* and *Dha* and rises to *tār ṙe* in the end. The melody drops back to *Ni* and *Dha* for the next word *ko* and the melody descends with the word *aṁkhiyā* sung to the phrase *D__M M\GrS*. The word *tarasa* that follows is broken into its three syllables, sung as a series of rising and falling trills—*Sa* rising to *Ma*, *Ma* falling to *Ga* and rising again and, lastly, *Ga* dropping to *re* and down to *Ḍha*. For *rahī*, it rises to *Ṇi* from *Ḍha* and then to *re* before going back, down to *Ḍha*. The melodic line now descends from *Sa* to *Ḍha* and rises *re* via *Ṇi* for *nā* while *jānu* is sung by ascending from *Ga* to *Dha* via *Ma* and ends on an extended *Dha*. *Kabā* is sung as a slight descent from *Dha* to *Ma* and back to *Dha* while *ghara* is sung as a *M\Gr__mīṇḍ* with the tone *re* extended. In last word *aveṅge* the initial vowel is sung first as a trill rising from *Ḍha* to *Dha Pa* in an undulating manner and descending to hold *re* followed by another section with ornamentation descending in the end to hold the *mandra Ḍha*.

After repeating the *mukhṛā*, faster *sargam tāns* follow with the development focusing on the upper tetrachord and above touching higher *ṙe*; the ascent, at times from *re* to *tār ṙe*, is followed by the characteristic descent in the *mīṇḍ M\Gr_* with an extended pause on *re*.

The melodic line thus arches upwards to *Dha*, at times beginning from the tonic but more often from *re* and descends but almost never to the tonic. The ascent to the higher tones *Dha* or *Ni* is either a sudden rise or in stages, taking each tone stepwise, but the descent is always gradual. The melody ascends to the high tonic, a few times, and holds it without descent to establish this key tone, with succeeding phrases following only after a gap. In the end, the melody returns to the lower tetrachord and the performance ends on the tonic with a cadential drop to the lower *Ṇi*.

2. *Khayāl (tīntāl)*

The second *bandiś* begins with *guru* sung to the tonic before dropping down to *Ṇi* for the syllable *bi* and ascends to *re* for *na and* commences *gyana* on this tone. It then drops to a prolonged *Ḍha* (via *Ṇi*) and ascends again to *re* and *Ga* then *Ma* for the second syllable *na* of *gyan* and the next word *na* and returning to *re*, prolonged for *pave*. The *mukhṛā* is now repeated.

Thereafter, the melody descends to the lower octave as far as *Ṃa* via *Ḍha* for singing *mana mūrakha soca* and ascends to *re* for singing the first syllable of the second *soca*. After a brief descent for *kāhe*, the melodic line ascends to Dha to sing *pachatāve*, with the *ta* sung to this extended tone, before descending from *Dha* to *Ḍha* via *Ga* and *re* and ascending to the tone *re* to end the last syllable *ve*. This is followed by the *mukhḍa* and *sargam* and other *tān*-s.

Thus, the melodic line in the *sthāyī* basically consists of an ascent from the tonic to *re* and descent to the lower *Ḍha* via a glide followed by an ascent to *Pa* and a descent to *re* again with a pause on this tone in the *mukhṛā*. In the second line, the ascent is stepwise, from the lower *Ḍha*, in two or three stages each time, reaching a higher tone first *Sa*, then *re* or *Dha* from where it descends to *re*. However, two features stand out in this faster *tīntāl* rendering, the first is that the avoidance of the tonic is not so stringent and the second is that the distinction between the *śuddha* and *komal* tones are not so strictly maintained as *komal ma* and *ga* and *śuddha* Re are used, albeit in passing.

In the *antarā* the melody rises much higher and touches the higher *Ṡa* and beyond to *ṙe* but the descent is similar with a glide or a quick drop to *re*. However here, too, we find that the tonic is sung more often along with a few instances of rest on the tonic.

To conclude from this examination of the two performances, the features specified in the texts, namely, the pause or prolongation of *re* in ascent is not manifest though *re* is intoned clearly from time to time. The higher *Ṡa* is often directly taken without descending to *Ni* and then ascending from *Dha*, despite this stipulation. The definitive tonal phrases indicated are not heard, as such, even in the *sargam* sections. However, similar phrases appear in the lower octave development. Thus, these 'definitive' phrases seem to be important for pedagogic reasons but get transformed in actual performance.

As far as the tonal hierarchy is concerned, it is that the tone *re* is prominent followed by *Dha* as stipulated. Next in the order of usage is *Ni* followed by *Ma* with *Ga* is least in terms of prominence. However, this by itself does not convey the image of Mārvā.

Summing up the performances, in our view the impression of Mārvā is conveyed by the steep rise from the lower octave, whether from the lower *Ḍha* or *Ṇi* to the *Dha* of the middle octave where the melodic line flattens because of the pause on *Dha* and this is followed by the glide down to *re* via *Ma* and *Ga*. This brings out the melodic opposition of *re* to *Dha* that contributes to the unique image of the *rāga*. This feature is not exhibited directly in the ascent as the tone *re* is usually not the initial tone for the ascent and nor is it paused on but is hinted at through the abrupt rise to the middle *Dha* from the lower tetra-chord, if not from the *mandra Ḍha*, and very infrequently from *re*, itself. This opposition is more emphatically conveyed through the descent to *re* from *Dha* via the glide, with the underlying tension deliberately not resolved by avoiding termination on the tonic.

In this melodic arch the descent is anticipated and a linear pattern of tension is developed with the expectation that the melody can only fall back to the pitch where it started. It is by denying this expectation that the *rāga* conveys its unique identity. Further in the *k͟hayāl* renditions of Mārvā, it is the sudden jump to the middle octave *Dha* that makes the descent important in conveying the form of the *rāga* as it fills the gap. This aspect of Mārvā is confirmed by various performers of the *rāga* and although Marwa is a rich and complex melodic entity this feature of the raga stands out [40]. In our view, it is the cue that registers in the immediate memory of the ordinary listener and functions to summon up the schema of the *rāga* Mārvā.

Performances of Pūriyā

Amir Khan (Navras NRCD 0092)

K͟hayāl tīntal

The first *bandiś* consists of only the *sthāyī* with the *mukhṛā China china bāṭa*(*ta*). The first two words are sung by rising from *Sa* to *Ma* and by descending from *Ga* back to *Sa* (via *re*). Next, *bāṭa* is sung as *S r* ^{N}*Ḍ*, while *takata* is sung using as *Ḍ Ṇ Ḍ*. *Hŭ* is broadly sung as *S r M D*, and *torī* is sung as a *mīnḍ M G r \S*. The first line is repeated before *kaba āvẽ more pyāre* is taken up. Here, *kaba* is sung ascending as *Ḍ Ṇ r* continuing the ascent for *āvẽ* sung as *G M D N*. For the next two words, *more pyāre*, the melody descends again in the phrase *M G r S*.

Sargam tān-s such as *Ḍ Ṇ r G M D N M G M G r S Ṇ* and *ṂḌṆr GM* and *ṂḌṆr S* are used where the descent from *Ma* to *Ṇi*, via *re* and *Sa* can be noticed. The movement is more of a glide than a swoop, since it is without any degree of suddenness. This can also be noted as the line goes from *Ma* to *Ṇi* while singing the word *tori*. On the other hand, *kaba āve* shows the ascent from *Ṇi* to *Ma* which is again gradual being via *re* and *Ga*. Further, the emphasis on *Ga* and *N*i by pausing and frequently using these tones is also evident. After the tonic, *Ga* and *Ni* are the most frequently used tones. However, *Ma* too follows close in usage and *re* is next in line, although it does not appear to be slightly higher in pitch than the *komal re*. However, when it comes to the issue of *mīṇḍs*, especially the stipulated *D*^{G}*M\G*, the situation is much more complex as the *mīṇḍs* themselves do not fit into this pattern [36, 41].

In this performance, the emphasis is on the tonic followed by *Ga* and *Ni*, the pivotal tones. Since *Ma* plays a role in the characteristic glides from *Ma* to *Ga* and *Ni* to *Ma* or the even more noticeable *Ṇi* to *Ma* ascent, it features next in terms of usage.

Bhimsen Joshi (ECLP1321 (1973)/EMI CD-PSLP 5022 Reissued as SAREGAMA CDNF 150,447).

1. *Khayāl vilambit ektāl*

The performance begins with the tonic being intoned followed by a descent to the *mandra Ṃa*. This overture remains in this region taking *Ḍha* and *Ṇi* before returning to *Sa* and then ascending to *re*, which is held, albeit with brief touches of *Ga*, before dropping to *Ṃa*. It plays around in this region with phrases such as *Ṇ Ḍ Ṇ*, *Ṇ S r* etc.

The *mukhṛā Pyārī de ga(ra)* is then sung with the characteristic sudden rise from *Ṇi* to *Ma* (*Ṇ/M*) evident in the first syllable of *pyāri* while the second syllable is sung to *Ga* followed by a descent via *re* to an extended *Sa*. Thus, this word is sung as an ascent from *Ṇi* to *Ma* and descent through the *mīṇḍ* via *Ga* to *Sa*. There is a drop to *Ṇi* to sing *de* while *gara* is sung as an ascent from *Ḍha* to *re* and then a descent to *Ṇi*. The word *lāgo* is sung by beginning on the *mandra Ṃa* and extending the vowel *ā* to ascend through *Ṇi* and *Ḍha* to *Sa* and ending on this tone with the last syllable extended. In *kāṭe nā* the first word sung as a rise from *Ṇi* to *Sa*, prolonged, before a brief descent to *Ṇi* and the second as ascending from here (via *re*) to rest on *Sa*. The next word *jāe* is sung dropping from the tonic to *Ṇi*. The next set of words *dukha ke* shows the melodic line rising from *Ṇi* to *Ma* (via *re*) in an upward glide rather than a swoop and for *dina* dropping from *Ma* to *Ga* and then *re*. *Piyā*, on the other hand, is sung as an undulating phrase, broadly *Ṇ S Ḍ r G S*.

In the *baṛhat* that follows, *Ṇi* and *Sa* are intoned for long with brief ornamentation using the tones below such as *Ḍha* and *Ṃa*. After a few repetitions of the *mukhṛā*, *re* and *Ga* are brought into the ambit through leisurely phrases such as *Ṇ r G* with *Ga* being sustained. Thereafter, *Ma* is reached and *M G_* phrases using words such as *lāgo* are sung and the rise from *Ṇi* to *Ma* and even from a lower octave *Ṃa* to the middle octave *Ma,* as well as descents from *Ma* to *Ṇi* can be heard. *Dha* is then fleetingly touched and the *D\G mīṇḍ* follows. The word *gara* is sung as *M/D* while the first syllable of *lāgo* is sung as *D\G*. Thereafter, the upper *Ṁa* and *Ḋha* are touched in a *tān* and *Ni* and the upper *Ṡa* approached. *Ni* is then held while singing *lāgo* and the *Ma-Ni* ascent heard. *Dukha ke* is then sung in fragmenting the words and syllables so as to bring out the rhythmic aspects. The higher *ṙe* is also touched from *Ni* skipping *Ṡa*. The higher *Ṡa* is reached directly from *Ma* and held. In this phase the use of *D\M* and *D\G mīṇḍ*-s is evident.

In the *antarā,* the melodic contour rises to the upper octave with the word *Sadāraṅga* broadly sung in steps, one for each of its syllables. The first syllable *Sa,* is sung to *Ga* (approached from *Ma* above), as is the second syllable *dā* which is sung to *Ma* approached from *Dha,* followed by a rise to *Ṡa* and then a drop to *Dha,* while the syllable *raṅ* begins with *Ṡa* drops fleetingly to *Ni* and then continues on an extended *Ṡa* with the last syllable *ga*. The next word *ḍholana* is sung to the upper tonic with the line rising quickly from *Ni. Miliye* begins with the first two syllables sung to a prolonged *Ṡa* before briefly rising to *ṙe* for the extended vowel *ī*. The last syllable is sung in an undulating manner alternating between *Dha* and *Ṇi* as *D N Ṡ D N ṙ N D* before dropping to *Ma*, held with a brief touch of *Dha*, before ending in *Ga*. The vowel *ye* is further extended as, after a pause, there is a rise from *Ṇi* to *Ma* and a descent to *Ga* with the melodic phrase ending with a rise to *Ma* and descent through *Ga* to *Sa*. Singing in *ākār* follows, interspersed with the first line of the *antarā* and variations on it before the last line *sukha nīta nāhī* is sung with *sukha* sung as *S M G* with the *Ga* extended for the last vowel while *nīta* is sung with the *ī* extended as an ascent with undulating phrase *D M Ṡ D Ṡ M G M*, *ta* is sung to *Ga* and *nāhī* is sung as a rise from *Ṇi* to *re* and back to *Sa*.

The development that follows consists of taking the upper register going up to *tār* ṙe and *tans*.

From this recording, the rise from *Ṇi* to *Ma* (*Ṇ/M*) and the drop from *Ma* to *Ṇi* is evident, both in the rendering of the *bandiś* as well as in the *baṛhat*, although this is often in the form of glides rather than a 'sudden' swoop juxtaposing these two tones as mentioned as an important feature. Other glides too are heard but it is not clear that it is the prescribed $D^{G}M\backslash G$ glide. In the *antarā* the tone *Ga* is heard in conjunction with *Dha* but it does not seem to be so frequent nor can it be identified as the *Ga-Dha* conjunction which is said to be an important feature of the *rāga*. Although, the upper tonic is taken from *Dha* or *Ma* in the *baṛhat* or while singing the *antarā*, there are some exceptions to this, especially in the *tan*-s. It would appear that this feature does not contribute in determining the identity of the *rāga* but follows from the requirement that *Ni* should be avoided if *Ṡa* is to be used as *ṇyāsa*. In terms of the frequency of the use of tones, the hierarchy is observed that after the tonic, in the case of the *vādi-samavādi* the latter *Ni* is sung more than *Ga* and besides these, *Ma* is the next in terms of usage. Although, *re* and *Dha* too figure, the use of these two tones is limited as laid down as a greater frequency would indicate shades of Mārvā. In this sense, the greater emphasis on *Ni*, the *samavādi*, rather than on *Ga*, the *vādi*, differentiates this performance by Bhimsen Joshi from the earlier one by Amir Khan.

2. *Khayāl drut tīntāl*

In the *tīntāl khayāl* that follows, the first word of the *mukhṛā ghaṛiye gi*(*nata*) is sung as a rise from *Ṇi* to *re* followed by a descent to *Ḍha*, a brief rise to *re* again and then to *Ṇi*. The next word, *ginata*, is sung as a descent from *Ṇi* to *Ḍha* and then to *Ṃa*. This comprises the *Ni-Ma* descent, but it does not register as a sharp change, as the tones are both from the same register. There are three repetitions of this phrase with the *Ṇ-Ḍ* and *Ṇ-Ṃ* tone juxtaposition before the next line, *tehārī milana kī āsa*, is taken up. For the first word there is a fleeting drop from the tonic to *Ṇi* before the ascending to *re* and return to *Sa* before resting on *Ṇi* for the vowel *ī*. The next word, *milana*, is sung as a slight ascent by dropping to *Ḍha* and then up to *Ṇi* with the last syllable is sung to *re*. The word *kī* is sung to Ga. The last word *āsa* is sung ornamented as an undulating line rising from *Ga* to *Ma* then to *Dha* before descending to *Ga* and rising to *Ma* again. From here the melody descends to Ṇi, thus, exhibiting the characteristic

M/Ṇ "swoop", before rising to *re* and then being resolved by a drop to an extended *Sa.*

The *antarā* that follows almost immediately is sung using tones in the upper half of the register with *ghaṛi* sung by taking a *Ma* from below for the first syllable and ascending to *Dha* for the second. The next few words are sung broadly around the *tār* Ṡa with *pala* sung to Ṡa and *china* commenced on Ṡa with fleeting drops to *Ni* before finally dropping to end on this tone. The next word *līna* sees a rise to the high *ṙe* for the first syllable and a drop to *Ṡa* for the second while the first half of *dhyāna* is sung as a rise to *Ṡa* from below and on to *ṙe* before descending to *Ni* and *Dha* for the second half. The word *hai* which is sung in conjunction with the previous one shows a trace of *Pa* as it drops to *Ma* from *Ga* and then rises to *Dha* before dropping to *Ga* and *terī* is basically sung as an arc rising from *Ga* to *ma* then descending to *re* and *Sa.* There is the ascent from *Ṃa* to *Ṇi* in *ginata* but it is not heard as a *mīṇḍ* connecting the two tones. The two tones *Dha* and *Ga* are heard together within complex *mīṇḍ*-s as in the word *āsa.*

The last line of the *antarā (jo mohe milana kī āsa)* is not sung immediately, but in a later cycle with *jo* being sung to *Sa,* and *mohe* sung by descending to *Ṇi* then rising to *re* before dropping to *Sa* and further to *Ṇi.* For *milana* there is a drop to *Ḍha* before ascending to *Ṇi* and *re.* To sing *kī, Ga* is held while *āsa* is sung as an undulating line rising from *Ga* to *Ma* to *Dha* (with slight drops in between) and then to *Ga* before to *Ṇi,* rising in between to *Pa*) in the course of this descent and finally rising to *re* to pause on this tone.

Thus, this recording does not add anything further to the conclusions that we arrived at from the analysis of the *baṛa k͟hayāl.*

From the detailed analysis of the melodic movements in the performances of the *rāga,* we can see that *Ni* and *Ga* are emphasized as required by the normative/prescriptive model and *re* is infrequently taken. However, these accounts are silent about the use of *Ma* which seems to play a significant role in the *chota k͟hayāl* performance, often sung *andolita.* This feature is not seen in the first two recordings that were analyzed.

As far as the stipulation of approaching the upper *Ṡa* avoiding *Ni* on the way up is concerned, since the performance was largely restricted to the middle (and lower) register, the issue does not arise. However, despite the dip to *Ṇi,* there is in Pūriyā, a frequent return to *Sa,* used as a rest-note. Thus, this feature itself differentiates the *bhāva* of Pūriyā from its parent Mārvā which avoids the tonic.

Leaving these details aside and focusing on how the listener would recognize a melody as Pūriyā, it is felt that the cue that assists the listener in doing so is the ascending movement of four intervals from the emphasized *Ni* to *Ga,* a consonance of perfect fourth followed by *Ma,* the fifth, another perfect interval before the drop. There is both oscillation of this tone and the use of *Dha* alongside *Ga.* These nuances provide character and structure to the melodic rise from *Ga* to *Ma* and back to *Ga* and provide an element of plaintiveness or seriousness or contemplative character, depending on one's point of view. Moreover, the drop back to lower *Ni* and resolution by stopping at the tonic makes this schema unique. It is this that provides the cues to the listener to differentiate Pūriyā from Mārvā rather than the issue of pausing on re in the ascent or noticing the extent of usage of *re* and Dha rather than *Ni* and *Ga.*

Performance of Sohanī

Malini Rajurkar (Fountain Music Co. FMCRD 053)

Khayāl drut tīn-tāl

Beginning with a quick *Sa* the short melodic phrase touches *Ma* to descend to *Ga.* Again from *Ga* it ascends to touch *Dha,* drops to *Ma,* touches *Ni,* then drops to *Dha* and gets back to *Ni held,* thus, *G__*^{D}*M* ^{N}*D N_D* followed by a phrase *M* ^{N}*Ś ___N \D__* where the high *Ṡa* is held sustained before a glide from *Ni* to *Dha.* The next phrase is a glide from *Ni* to *Ga* (via *Ma*) and an undulating rise to *Ma,* return to *Ga* and *Ma* again, before dropping to the tonic, thus, '*D\G__M G_MG S_*'. In the last phrase, there is a stepwise rise from *Ga* to *Ṡa* (via *Ma* and *Dha*) and the *tār Ṡa* is then held with brief drops to *Ni* and *Ma.* Thus, in the *avachar* after beginning with *Ga* the melody ascends—with the main movement around the high *Ṡa*—before it descends via *Ni, Dha, Ma* and *Ga* to the lower *Sa* and then returns to the upper tetrachord.

The first word of the *mukhṛā, kāhe,* is sung to *Ṡa* and drops from there to *Ni* for the second syllable of the word *aba* and to *Ma* and *Dha* for *tuma* before returning to Ṡa, now held in extended singing of the vowel *ā* and then gliding to *Ni* and *Dha* for the next syllable *ye* and the next word *ho.* The *mukhṛā* is repeated broadly following the initial melodic pattern. In the next set of words *mere dvare,* where the first word is sung to *M*^{G}*MG* while dropping to the lower *Sa* for the second. The second line *sautana sanga jāge* is sung to the lower *Sa* for *sautana* and rises to *GMG* for *sanga jāge* continuing as *M G M D S__* for *rasa pāge* and *MGGM* for *anurāge pāge.*

The *baṛhat* begins after repeating the *mukhṛā* twice. This development is in *ākār* and largely in the upper-tetrachord with *tān*-s such as *G M D M, G M D M D M D M* or *M D N D* interspersed with the *mukhṛā* as a refrain, small *ākār tān*s and variations of the *mukhrā* and prolonged intonation of *Dha* taken from *Ga,* sometimes ascending from *Ga* to reach the *tār* Ṡa, the tone being ornamented by quick dips to the tones immediately below. Thus, the development is almost wholly in the region around the high *Ṡa* although *ṙe* is not taken. The ornaments that are used are *D\G mīṇḍ*s and rising undulating phrases rising from *Ga* to *Ṡa* are noticeable when the *mukhrā* is varied and also when the word *āye* is sung.

The first line of *antarā* is sung after *baṛhat* rising from *Ma* through *Dha* and *Ni* to *Ṡa* and holding the upper tonic. In the second repetition of the *antarā ṙe* is touched followed by similar fleeting touches and it is only intoned in a sustained manner later in *ākār* with a touch of the *tār Ĝa.* The line is repeated, melodically varying it. Thereafter the *mukhṛā* is repeated with variations along with *tān*-s.

This is followed by a *tarānā,* which we do not analyze.

In Sohanī, *tār Sa* is directly approached via *Ni*, unlike Mārvā and Pūriyā. While both Pūriyā and Sohanī have the tone combination *D G M G*, in Sohanī this combination is approached in descent from the tone *Ni*. In Sohanī, there is very little possibility of confusion between it and the other two *rāga*-s because the melody persists in the upper tetrachord. But mere registral movement should not be the criteria for distinction and this *rāga* which has its identity indicated by the rise from *Ga* to the upper *Ṡa* and the consonance of a perfect fifth between these two tones. The melody also tends to hover around the high *Ṡa* where embellishments such as the *andolan* and *ṙe* used as a grace tone are employed to relieve the lack of variety that sticking to this high pitch entails. Oscillating *Ni* and *Dha*, during descent, can be regarded as having a similar purpose. While the drop to *Ma* and then *Ga* achieves closure by returning to what is basically the starting note.

To sum up this performance, Sohanī's salient feature is its quick rise to the upper tetrachord and its movement in this region. It is, thus, regarded as a 'light' *rāga* both in terms of its mien and also because of the limited possibilities of melodic development through variation. Its rules are flexible as the use of accidentals, *komal dha* and *shudha Re*, is permitted and it is therefore largely used for genres such as *ṭhumrī*.

To conclude, after analyzing the recordings of the performances of the three *rāgas* Mārvā, Pūriyā and Sohanī, it is observed that the ascent and descent patterns and tonal phrases that are usually mentioned in the authoritative texts that list the features of these *rāgas* are not heard in their actual performances. Such phrases and *chalans* might be important to the artiste and performer in imbibing the *rāga* and instantiating it, but do not help the listener in identifying the *rāga*. Instead, the three *rāgas* can be recognized by certain cues which are not necessarily full tonal phrases but are unique features such as the descent from *Dha* to *re* in the case of Mārvā without closure on the tonic. In the case of Pūriyā, it is the consonance of the lower *Ṇi* and the rise from here to *Ga*, a consonant fourth and further on to *Ma* with a sudden drop from this tone back to the *Ṇi* that constitutes its uniqueness and the cue that helps in its differentiation. Finally, as regards Sohanī, it is the melodic rise from *Ga* to the upper *Ṡa*, the persistence of the melody in these regions and the drop to Ni.

Bhūpālī and Deśkār

The Prescriptive

Bhūpālī and Deśkār are both pentatonic *rāga-s* with the fourth (*Ma*) and sixth (*Ni*) tones omitted, and thus share the same tonal material. In Bhūpālī, *Ga* and *Dha* are important notes with *Ga* being the focus of melodic development while *Sa* and *Pa* are also significant as they function as rest tones. The melodic movement of Bhūpālī is mainly in the lower register and middle registers whereas for Deśkār it is usually in the upper tetrachord, commencing from *Pa*.

The initial phrase in the case of Bhūpālī is *S Ḍ S R G*_ and rest is permitted on all four tones: *Sa*, *Re*, *Ga* and *Pa*. Four sets of definitive tonal sentences for the different regions of the scale are indicated for Bhūpālī with the tonal activity around *Ga*, the *vadi*, and includes the *G-D* coupling, the linking of tones through glides, and the emphasis on *Pa* during ascent. The *rāga* unfolds in the lower tetrachord with the phrases *S R S* ^{s}*Ḍ S* or *S R* ^{s}*Ḍ S* are used to mark the end of the development [41]. Glides between *Pa* and *Ga* (*P\G*) and *Sa* and lower *Ḍha* (*S\Ḍ*) characterize the melodic movements of the *rāga* with the likelihood of *Ma* and *Ni* being heard in these glides. Some regard the *rāga*'s ascent and descent as non-linear while others view them as straightforward except for the specific feature where the tones *Re*, *Ga* and *Dha* are usually approached from above in the ascending phase, while during descent, the latter two are linked by glides [42].

In Deśkār, on the other hand, *Pa* and *Dha* from the upper tetrachord are important, alongside the *tār* Ṡa. The definitive tonal 'sentences' for Deśkār are mostly in the upper tetrachord and where the higher *Ṡa* is taken. *Pa* is used as a rest only while descending and *Re* is weak or is skipped, whereas it functions as a rest tone in Bhūpālī. The *vādi-samavādi* pair is *Dha-Ga* for Deśkār as against *Ga-Dha* in Bhūpālī. However, it is contended that *Dha-Ga* conjunction in Deśkar is unique in the manner of its coupling, as *Ga* is approached from either *Dha* or *Pa* above, while in Bhūpālī, *Dha* is approached from *Ga* below [41]. Other sources state that while the ascent descent may be the same for both, in Deśkār both the tone treatment and the movement is different as *Pa* and *Dha* are both sustained, with a slight oscillation on the latter, and takes place in the upper tetrachord and the higher register above the upper tonic. The *rāga*'s characteristic movements are shown as Ṡ_ ^{RG}*P*_ *D*~ *P*, *G P* *D* *D* /*S*, ^{DDPG}*P*_ ^{GPD}*P G*_ *R S* [42].

Analysis of the Performances of Bhūpalī and Deśkār

For understanding Bhūpalī, two recordings have been selected, one each by D.V. Paluskar and Kishori Amonkar, both commercially released and available. As far as Deśkār is concerned, the commercial recording by Mallikarjun Mansur, also transcribed by Magriel, is available and has been analyzed alongside.

Performances of Bhūpālī

1. D.V. Paluskar (CBS Swarashree/ AIR DV001 1988)

Khayāl vilambit ektāl:

The *bandiś* begins with a short melodic introduction with quick *Sa* followed by a rise through *Re* to *Ga*, a pause, before a drop to *Re* and return to *Ga*, now held extended. After this, there is a drop to the tonic and a rise to *Re* ending in a descent to *Ḍha* followed by the tonic. This phrase, *S R Ḍ S_*, is one of the characteristic ones marking the end of melodic development.

The *bandiś*, *"Jaba hi saba nirapata"*, begins with an ornamental flourish, followed by the word *jaba* sung broadly as *G R S_* with a drop to *Ḍha*, the glide short but evident; *hi* is then sung to an extended *Sa* ending with a glide to the lower *Ḍha* (*S\Ḍ*). S*aba* is sung as a rise from *Sa* to *Ga* ending with a *G\R mīṇḍ* on the second syllable. The word *nirapata* is sung extended with the first syllable sung as extended *mīṇḍ P\G* with a pause on both the tones, followed by a quick rise to *Pa* and then a short *G\R* glide. The second syllable *ra* is sung extended to *Ga* while *pa*, the next, begins with an ornament followed by a glide from *Pa* to *Ga* while *ta* is sung, basically, as a *G\R mīṇḍ* ending on the tonic. The *mukhṛā* is then repeated to the same melodic outline followed by the word *nirapata* which is again sung broken into syllables melodically similar to its earlier utterance. Thereafter, in *nirāsa*, the *ni* is intoned to *Sa* and *ra* too is continued on the same tone before rising to *Re*. In *bhaye* which follows, *bha*, the first syllable, is sung as a rise from the lower octave *Ḍha/ṇi* to *Pa* and thereafter the syllable *ye* is sung through a *mīṇḍ P\G* and further extended through *G\R_S_ mīṇḍ* with a flourish introduced in between, before again continuing the glide. After a brief trill, the *mukhṛā* is sung again and variations in *ākār* are then commenced.

In the first stage of the *baṛhat* that follows, the region between *Ga* and *mandra P.a* and *Ḍha* is developed. The tones *Ga*, *Re* and *Sa* predominate with *Ga* and *Sa* intoned in an extended manner linked by descending *mīṇḍ*s, either *G\S* or *G\R S_*, or as *S/R* or *S/R\S* in ascent. The ascent from *Re* to *Ga* via a *mīṇḍ* can also be clearly heard. There are a few instances of these tones being approached from the lower octave usually *Ḍha* and in one instance even from *P.a*. The middle *Pa* appears at the end of an ascending *mīṇḍ* or at the beginning of a descending *mīṇḍ*. Later, it is heard frequently at the beginning of the descending *mīṇḍ* and after *Dha* is reached, it appears in the phrase *DP\G*. A noticeable feature here is the step-wise ascent to *Re*, *Ga*, and *Dha* usually approached from above. In such ascents, *Ma* and *Ni*, the two tones prescribed in the texts as to be omitted, are nevertheless, occasionally heard here. The upper tonic *Ṡa* is touched shortly thereafter. A similar step-wise ascent from *Pa* to *Ṡa* follows and the descents are through glides. In this last phase, *Ṡa* is often sustained and the upper *Ṙe* is sung followed by the descent to *Ṡa* in a glide. The melody also touches the upper *gandhār*. The descent from *Ṡa* to *Sa* is often in a series *mīṇḍ*-s via the intermediate tones. And this pattern of ascent and descent provides the structure for the *ākār tān*-s that follow the *bol banṭ* section and also the short ascending or descending *tān*-s and *phirat tān*-s.

The *antarā* comes almost towards the end of the performance as the word *Guru* is sung basically as a quick *P\G mīṇḍ* while for the syllable *pa*, the tone *Pa* is held and *da* is sung as a glide from *Ṡa* to *Dha* (*Ṡ\D*). The next word *kamala* is sung extended, with the first syllable sung as an ascent from *Pa* to *Ṡa* and a pause on the tone for the other two syllables, *ma* as well as *la*; *vande* that follows, is also sung to the upper tonic with a brief drop in between to *Ni* before ascending via a glide to the upper *Ṙe* and terminates on this tone. *Raghu*, too, is sung to the upper tonic with a brief drop to *Ni* before ending in the upper *Ṙe*, while for *pata* the syllable *pa* is sung to Ṡa and descending to *Dha* in a glide for *ta,* with the terminal vowel *a* being extended and sung to *Pa,* to merge with the next word. *Taba* begins with the first syllable being sung to *Pa* and descends to *Ga* for the second syllable *ba* with its end vowel extended and sung by descending to a prolonged *Sa*. In *chāpa*, the first syllable is sung by descending from *Pa* to *Ga* and for the second syllable, by descending from *Ṡa* to *Dha*. In singing *sāmīpa*, the first syllable is sung by ascending from *Pa* to *Ṡa* and pausing on it, and then briefly dropping to *Dha* before returning to the upper tonic for the next syllable. For *mi* there is a brief drop to *Dha* before returning to *Ṡa,* with the extended vowel *ī* sung ornamented, before ascending to *Ṙe*. For the next syllable *pa*, there is a rise from *Pa* to *Ṡa* before dropping to *Dha*. The first syllable of the word *gaye* is then sung as *D_P* while *ye* is a descending *mīṇḍ* from *Pa* to *Ga* (*P\G*) and then a brief descent to *Re*. The syllable is further extended by rising from *Re* to *Pa* and then descending to *Ga* and rising to *Dha* before descending to *Ga*, each one of these movements being through glides. Thus, in the *antarā*, the lines are sung to a series of *mīṇḍ*-s, the most prominent of which are *Ṡ/Ṙ*, *P\G*, *P/Ṡ*,*Ṡ\N*, *Ṡ\D* before its second line returns to the lower tetrachord with *G\R*, etc and joins up with the *mukhṛā*.

All tones of this pentatonic *rāga* are deemed to be important tones: *Sa*, *Re*, *Ga*, *Pa* are those where the rest is permitted and *Dha* is regarded as the *samavādi*. However, in this performance, while this hierarchy is by-and-large observed, and after the tonic, *Ga* the *vādi* is followed by *Pa* and then *Re* in terms of frequency of usage. However, *Dha*, despite being the *samavādi, is the* least intoned. In shaping this *rāga*, it is seen that the *mīṇḍ*-s, such as the descending *G\R, P\G* and *G\S*, play an important role. Alongside these, the ascending glides *R/G* and *D/Ṡ* are also present but to a lesser extent. The emphasis on *Pa* which is often at the end of an ascending movement forms an important feature. The feature where the tones *Re*, *Ga* and *Dha* are approached from above is also noticed.

2. Kishori Amonkar (HMV ECSD 2702(1972)/EMI CD PMLP 5816:

K͟hayāl in vilambit tīn-tāl

The *k͟hayāl* starts with a short melodic outline with an extended *Re*, a gradual rise from *Ḍha* to the tonic and onwards through *Re* to *Ga* in a gentle graceful *miṇḍ* indicative of the *rāga*. This is followed by an undulating melodic phrase where *Ga* is taken in ascent before dropping to *Re* and back to *Ga* followed by another drop before rising to *Pa*, a feature of the *rāga* where tones are approached from above. This melodic piece ends with a descent to *Ga* and *Re* through a glide. The melody rises to *Ga* and after a return to *Re*, briefly drops to *Ṇi*, before rising to the tonic. This kind of a movement continues in which *Re* and *Ga*, more particularly the latter, are emphasized in an ascent from the tonic and descent back. This prelude ends with *Pa*, approached through an undulating phrase, and the tone is held sustained before dropping to *Ga* with the characteristic *P/G mīṇḍ*.

The *bandiś* commences with the first word *erī* where the first syllable is sung extended rising from the tonic to *Ga* before dropping back in a *G\S* glide and then rising to *Re* before returning to a prolonged *Sa* for the second syllable. In *āja*, the first syllable is sung with the vowel extended by a rising to a prolonged *Pa* from *Re* and then descending in a deliberate and discernible glide to *Ga*, and pausing on this tone, before quickly ending on *Re*. Similarly, the second syllable *ja* is sung with the vowel extended holding *Ga* and then *Re*, the tones linked by a *mīṇḍ*. The next word, *bhaīlavā*, is similarly treated with the vowels prolonged and syllables sung linked by *mīṇḍ*-s. While its first syllable is sung to the tonic, (approached from *Ḍha*), the vowel *ai* is sung to an extended *Re*; the second syllable, *la*, is sung relatively quickly to the tonic and the last, *vā*, is sung with the vowel extended gliding from *Re* to a prolonged *Sa*. For *sukhavā*, there is a gradual descent from *Sa* to *Ḍha* for the first syllable, *su*, and an ascent to *Sa* followed by an ascent to *Re* for the second, *kha*, as the end vowel is stretched. Similarly, the last syllable, *vā*, commences with a brief ornamentation before touching *Ga* and the terminal vowel is then sung by dropping to *Re* in a glide and ascending again to *Ga* to end in *Re*. For the next word, *more*, the first half is sung with a brief ornamentation and ascent to *Ga* and a descent to the tonic via *Re* while the tonic is held for the second half. Then *jiyā* is sung with the first half as a glide from *Ga* to *Pa*, held for long on the vowel, before a trill that leads on to the next half after which the melody returns to *Pa* before descending to *Ga* and then to *Re* in a glide. Next *ke* is sung, again basically, holding *Ga* and ending in a glide to *Re*.

The next line then commences with the word *suna* repeated twice, the first time as *P\G/P*, the second time with a short-ornamented beginning rising briefly to the upper tonic and then descending to *Dha* and then back to *Ṡa* held for the extended vowel. This is followed by *piyā* where the first syllable is sung rising from *Pa* to *Ṡa* followed by a brief drop to *Dha* and then back to *Ṡa* with a glide back to *Dha* and continues on this tone to sing the next half as a *Dha-Pa* glide. Thereafter, *kī* is sung as a *Pa\Ga* glide with the last tone extended for the vowel. The last word of the *sthāyī* is *bāta* where the first half is sung with a brief ornamentation beginning with *Sa* and rising to *Ga* dropping back to *Sa* and then onto *Ma* and dropping back to *Ga* rising to *Pa* and then dropping to *Ma* to rise to *Dha* and hold it. Here we can hear *Re* and *Ga being* approached from above. We can also hear *Ma*, a tone that is not part of the scale being intoned. The last syllable is then sung as a *P\G\R* glide before returning to *erī*, the first word.

The *antarā avana kahilavā begi milā* follows after a short sarangi interlude. For the first word, the melody rises from a fleeting touch of *Ma* to *Pa* and then gradually slopes to *Ga* for the initial vowel and then rises to *Pa* and back to *Ga* with a brief drop to *Ma* for the syllable *va*. It drops to *Ma* and returns to *Pa* at the end of the syllable. The next syllable *na* is sung rising from *Pa* to *Ṡa* and then gradually drops to *Dha,* rising again to *Ṡa* in a slow ascending glide, with *Ṡa* paused on to sing the vowel which is extended. The first syllable to the next word *kahilavā*, is begun rising from *Dha* to *Ṡa*, dropping briefly to *Ni*, and back to *Ṡa* which is held and then *hi* sung with a fleeting drop before holding the tonic, albeit with a few drops to *Ni*. The third syllable *la* is sung rising from *Ni* to *Ṙe* and dropping back to the upper tonic to sing the last syllable. The next word *begi* is sung by approaching *Ni* from below then dropping to *Dha* and rising to *Ṡa* in a glide. For *mila*, the first syllable is sung by approaching *Ṡa* from below, through *Pa,* and then dropping to *Ni* before beginning *la* on the tonic which is held (barring a few fleeting dips to *Ni*), followed by a rise to *Ṙe* to descending to *Ṡa* to sing the end vowel. The first half of Sadaranga, the eponymous composer's name, is sung with the first half as a rise from *Ga* to *Pa* and a glide back to *Ga*. The next half sounds like *rahoge* and is sung ornamented as a series of ascents followed by descents from *Ga* to *Re* and back, continuing the rise via *Pa* and *Dha* to the *tār Ṡa*, and continuing to the upper *Ṙe* to touching the upper *Ĝa* and return in a glide to *Ṙe*, now held. In *anata* the first vowel is extended and sung as a undulating melodic line rising from *Pa* to *Ṡa* then onto *Ĝa*, dropping to *Ṙe* and back to *Ĝa* before an almost abrupt drop to *Ṡa*. *Na* is sung in continuation to *Ṡa* before a gradual descent to *Dha* and a drop to *Pa*. This is followed by *ta* which is sung with the terminal vowel extended as a *Dha\Pa* glide. The first syllable of *bhaīlavā* is sung by ascending from *Ga* to *Pa* and then to the upper *Ṡa* with the *ī* is sung extended to the tonic and the melody, thereafter, drops to *Dha* to continue with *la* and ends on *va* with a *Pa\Ga mīṇḍ*.

The melody having descended to the lower tetrachord, the last two words, *more kāja* are sung using the tones in this region with *mo* being sung rising from *Re* to *Ga* and back to *Re* while *re* is sung in continuation. In *kāja* there is the characteristic undulating rise from *Sa* via *Re* and *Ga* to *Pa* and then to *Dha* for the first half, followed by a descent from *Pa* to *Ga* and then *Re* for the second half. The *antarā* now returns to the initial words of the *mukhṛā,* albeit sung with slightly different tones.

The use of *mīṇḍ*-s is a notable feature of Bhūpālī and a crucial element in the identification of the rāga. *Mīṇḍ*-s are used while singing a single syllable of the lyric text and are identified with reference to these syllables. Often, in singing a word of two syllables, the second is extended and sung as a descending *mīṇḍ G\S* followed by another *S/R* and back (*R\S*). For instance, while singing the word *āja*, the first syllable is once again extended and sung to a sustained *Pa* with a glide *P\G* while the second is sung as *G_\R*. Similarly, *bhailavā* is sung by ascending from the lower *Ḍha* to *Sa*, the syllable *ī* is sung as *R\S*, while *vā* is sung to the *mīṇ-ḍ R\Ḍ*. In the case of *sukhavā*, there is an ascending glide from the lower *Ḍha* to the tonic. Instead of continuing with these examples, it would not be wrong to state that the rendering of the *rāga* is this performance is replete with *mīṇḍ*-s, much more than is evident in the previous case. In addition, the subsequent development in *ākār* as well as through *tān*-s also brings out the differences in styles.

As far as emphasis and usage of the tones is concerned, after the tonic which predominates, use of and pause on the tones *Re* and *Ga* is pronounced. As compared to the earlier performance of Bhūpālī, *Dha* as *samavādi* follows these two tones in the extent of emphasis and, thus, besides the difference in the extent of *mīṇḍ*-s, there is also a difference in the extent of emphasis on the tones.

Performance of Deśkar

Mallikarjun Mansur (Inreco 2411 5096 1981)

Khayāl in vilambit tīntāl

The performance begins with a short outline consisting of three melodic phrases. The first begins with an undulating melodic line where an ascent from *Sa* to *Ga* is followed by a drop to *Re*, a rise to the flattened *dha* then a drop to *Ga* tone before attaining *Dha* and holding it, with a slight oscillation. In the second, while the tone *Ga* is still held, there are a series of upward movements fleetingly touching the tones, *Ṡa* and *Ṙe*, above before descending to *Pa* and then *Ga*. In the last, *Pa* is held before being briefly oscillated, followed by a descent to the tones, *Ma* and *Ga* below.

The *sthāyī* now begins with *huṁ* sung to *Pa* while *to* is sung to *Dha* and the first syllable of *tore* is sung to *Dha* and rises for the second one which is sung to the upper *Ṡa*. In the next word *kāraṇa*, the first syllable is sung with the terminal vowel stretched, basically, to *Ṡa* ornamented by a couple of fleeting drops to *Dha* and *Pa* before fully descending to *Dha* and then *Pa* for the next two syllables, also with their end vowels extended. The word *jāge* is sung in continuation as an extension of the earlier one and ends in the middle *Sa*. In the second line, *so* is now sung conjoint with the earlier word by ascending from *Sa* to *Re* and this ascending line is continued as *balama* to Ga and the first syllable of *more* is sung as a descent from *Ga* to *Sa* (with a fleeting touch of Re in between) and the vowel stretched with the *re* terminating on *Sa*. From here, the melody descends to the lower octave *Ḍha* for *jāgata* and rises back to *Sa* at the last syllable *ta* with *bhailī*, the next word, sung ascending from *Sa* to *Re* and *Ga* and by the time the last syllable is sung the melodic line has ascended to *Pa* which is held while singing the end vowel. The last word of the *sthāyī bhore* is sung ascending from *Pa* to *Dha* (with a brief drop to *Ga*) which is held with a slight oscillation and its middle vowel is extended as the line drops to *Pa* then *Ga* and *Re* before returning to *Pa*. After a brief pause, the vowel is continued to be sung, now as an undulating contour oscillating between Dha and Ṡa. After another further pause, the *o* is continued to be sung through an oscillating descent from *Pa* to *Re* followed by an ascent back to *Pa* and then again, a descent to *Re* and *Sa*.

The presentation of the *sthāyī* is followed by the *baṛhat* which commences with variations in *ākār*. After returning to the *mukhṛā* this is followed by another round of variations, this time using the text of the lyrics. The next stage of development is in the higher tertrachord. The *mukhṛā* is repeated and then varied. The words *jāge so balamā* are then sung. There is a session of *baṛhat* after this, as variations in *ākār* and some *tān*s follow.

Almost nine minutes into the presentation, the *antarā* is presented. The first line begins with the word *aṅkhiyā* whose first syllable is sung rising from *Ga* to *Pa* and the second from *Pa* to *Dha*, with the last *yā* sung extended to a prolonged higher Ṡa the word *morī* is sung to the same tone but after a brief dip to Dha with the vowel of its first syllable stretched. There is a quick drop from *Ṡa* to *Dha* while singing *tuma,* but *saǜ* is sung with a glide rising from *Ṡa* to *Ṙe* and back while *lāge* is sung as a *mīṇḍ* from *Ṡa* via *Dha* to *Pa* and this tone continued for *ra* as the second syllable of this word is not sung. The words *aṅkhiyā morī* are repeated to the same broad melodic outline except that the last syllable of *aṅkhiyā* is sung ornamented and the word ends in a trill. The line is again repeated in a similar manner rising from *Ga* to *Pa* to *Dha* and then to *Ṡa* for the last syllable which terminates without the ornamentation but with a drop to *Dha*. It continues with a return to the tonic as *mo* is sung to it and *ri* sung with a drop to *Dha.* The melodic line is repeated by the supporting voice. For the next two words *tuma saŭ*, the first is sung as a drop from *Ṡa* to *Dha* while the second as *Ṡa* rising to *Ṙe* (with a brief drop to *Dha* in between) and then dropping back to the upper tonic for *lā* which ends on *Dha* and *ga* is sung continuing the descent to *Pa* and for *ta* the line descends to *Ga.* Here the text is modified as it is sung as *lāgata lāga rahī.* For the second *lā* it rises from *Pa* to *Dha* and drops again to *Pa* as *rahi* is sung in continuation with *ra* sung to *Pa* and the *hi* to *Dha* and for the last two words *cahū ora*, the first is sung, beginning with a quick drop to *Pa*, as *Dha* and the last word is sung as a trill. The *mukhṛā* is then repeated. After another delivery of the *mukhṛā*, the *baṛhat* begins with variations in *ākār* and *ākār tān*s as well *phirat tān*s and some rhythmic play using *bolbāṇṭ.* The main point is that while the structure of the *tān*s varies as does the length of the session, the *tān*s themselves do not descend below the middle *Ga.*

To sum up the performances of these two *rāga*-s, Bhūpālī and Deśkar, the first point that comes to notice is that their melodic progressions are different, as they occupy different regions of the register. While Bhūpālī is developed basically in the *mandrā* and the *Madhya saptak*, Deśkar jumps up to the upper tetrachord and moves beyond to the higher register where the *tār Ṡa* is sustained. Thus, this itself constitutes a significant difference and even though the basic tonal material might be the same, this changes the nature of the Deśkar's melody which gives prominence to the tones *Pa* and *Dha* with the latter tone being gently oscillated. Besides, the step-wise movement in Bhūpālī and its descent in glides makes this its distinguishing feature. While there is a hint of a glide from *Dha* to *Ga* in Deśkar, it does not have the kind of impact that the *mīṇḍ*-s in Bhūpālī have. Thus, while the presence of these *mīṇḍ*-s in descent and the gradual stepwise ascent in Bhūpālī characterize it and aid in its recognition, in the case of Deśkar, it is its sudden rise to the upper tetrachord and beyond and its melodic movement in the upper reaches that help in identifying the *rāga* and differentiating it.

Conclusion

We have examined certain aspects of the *rāga*, a melodic entity that is fundamental to the music of the sub-continent with a view to understand how a particular *rāga* is identified and cognitively recognized by a listener. In order to do so, we took two sets of *rāga*s, where the *rāga*s in each set shared the same tonal material and examined each set in an attempt to understand what were the differentiating features of the *rāga*s comprising each set. At first, the structural characteristics of each one of these, as laid down in texts commonly accepted as reliable and authoritative were outlined to understand their main features which were required to be established in the instantiation of that particular *rāga* through performance. Thereafter, performances of the same *rāga* as contained in commercially available recordings were analyzed using pitch-extraction soft-ware and the resulting melodic contours were studied to see which features were salient in performances and those that were unique to that particular *rāga* and functioned as a cue to the listener in the identification.

Based on this analysis, it can be concluded that the features of the *rāga* as prescribed for being observed in performances are pedagogical devices to help the musician understand the *rāga* and do not appear to be stringently followed in the process of performance. Moreover, the features to be emphasized or avoided to differentiate similar *rāga*-s sharing the same tonal material, as laid down in the texts, also appeared to be theoretical distinctions, and were not fully observed in the performances. It was noted that ascent and descent patterns by themselves were not adequate for *rāga* identification or differentiation in performance, although differences in the parts of the register where the melody was mainly sung was an important factor for distinguishing similar *rāga*-s. Further the motivic phrases (*pakad*) and melodic movements (*chalans*) said to characterize the *rāga* do not appear to be that significant in the process of identification and recognition.

On the other hand, it is the presence of certain cues that are unique to each of the *rāgas* that appear to aid in the listeners' process of differentiation and identification. These cues are small features such as ornaments, mainly glides and oscillations, and predominance of certain tones. The cues here correspond to "very restricted entity….often shorter than the group itself but always embodying striking attributes" [43]. These cues signify longer entities or groups such as motifs or phrases or even characteristic melodic movements. While this concept and the process through which these cues work have been proposed in the context of western music, these are equally applicable to Indian *rāga* music and the cognition of this entity. It is felt that such cues enable the recall of the underlying schema by the listener from memory. Once the schema comes to the listener's consciousness, it is elaborated and further developed by incorporating the other melodic phrases. In this sense, the cue is a partial instantiation of one striking feature of a schema which the listener selects from among several likely schemata. And as other features of this schema are instantiated, the listener is able to categorize the music and identify the *rāga* being performed.

This paper is a preliminary exploration into the cognitive aspects of the *rāga*, a fundamental concept underlying Indian art music, which is normally defined in terms of its structure and features. It is necessary to further investigate the role that the concept of a cue plays in *rāga* identification and the process of *rāga* categorization by studying other *rāga*-s and their performance, to identify the different types of cues that might operate in the case of a particular *rāga*. Such an effort will lead to a better understanding of the cognitive aspects of its nature and underlying process of categorization.

Notation Employed

In Indian music, the seven basic tones or *svara* are named as *śadja, riśabh, gandhar, madhyam, pancham, dhaivat and nishad*. In usage, they are abbreviated to the syllables *Sa, Re, Ga, Ma, Pa, Dha*, Ni, similar to the solfege symbols of the western system. Except for the tonic, *Sa* and the fifth *Pa*, the remaining five svara-s can be altered from their natural position. *Sa, Re, Ga, Ni, Dha* can be lowered by a semitone and these variants are referred to as *komal*. In the system of notation adopted here, the lower case letters represent the *komal* forms. In the case of *madhyam*, its variant a semitone above, is termed the *tīvra*, represented here by the upper case *Ma* while its natural or the *shuddha* form is shown in the lower case as *ma*. All these twelve tones are shown in their abbreviated form, thus:

S, r, R, g, G, m, M, P, d, n N, Ṡ

The three registers or octaves are termed *mandra* (lower), *madhya* (middle) and *tār* (high) and indicated by a dot below for the low (.) and a dot(•) or dash(') above for the high.

Although there are many ornaments and embellishments that are used in Indian music, for the sake of simplicity only two have been shown here. The *mīṇḍ* or glissando with the ascending glide shown as '/' and the descending one as '\'. The *andolan* or oscillation has been shown as '~'.

Lastly, proportional notation has been used to indicate relative and approximate duration of tones with a horizontal line '__' indicating the pause or *nyasa* and tones being placed closer to each other indicating faster passages.

Acknowledgements I am grateful to Kaustuv Kanti Ganguli for introducing me to both the Melodia plug-in and Sonic Visualizer and in providing me with his script "Audio signal processing for melodic analysis of Indian art music". Nandan Bagchee helped in customizing this for the purposes of analysis in this paper.

References

1. W.B. Hewllet, E. Selfridge-Field, Computing in musicology, 1966–91. Comput. Human. **25**, 381–392 (1991)
2. D. Meredith (ed.), *Computational Music Analysis* (Springer, 2015)
3. K. Gajjar, M. Patel, Computational musicology for raga analysis in Indian classical music: a critical review. Int. J. Comput. Appl. **172**, 9 (2017)
4. S. Shetty, K.K. Acharya, Raga mining of Indian music by extracting Arohana-Avarohana pattern. Int. J. Recent Trends Eng. **1**(1) (2009)
5. G.K. Koduri, P. Rao, S. Gulati, A survey of raaga recognition techniques and improvements to the state-of-the-art. Sound Music Comput. (2011)
6. G.K. Koduri, S. Gulati, P. Rao, X. Serra, Raga recognition based on pitch distribution methods. J. New Music Res. **41**, 4 (2012)
7. P. Rao, J.C. Ross, K.K. Ganguli, Distinguishing raga-specific intonation of phrases with audio analysis. J. ITC Sangeet Res. Acad. **58**, 26–27 (2013)
8. J.C. Ross, T.P. Vinutha, P. Rao, Detecting melodic motifs from audio for Hindustani classical music, in *Proceedings of International Conference on Music Information Retrieval* (2013), pp. 499–504
9. S. Gulati, J. Serra, K.K. Ganguli, X. Serra, Landmark detection in Hindustani musical melodies, in *Proceedings of the ICMC 2014, Athens, Greece* (2014)
10. P.V.G.D. Prasad Reddy, B. Tarakeshwara Rao, K.R. Sudha, C.V.M.H.K. Hari, Automatic Raaga recognition system for Carnatic music using hidden Markov model. Global J. Comput. Sci. Technol. **11**, 22 (2011)

11. T.P. Vinutha, Music structure analysis of Hindustani music for transcription, in *International Journal of Current Engineering and Technology/Proceedings of the National Conference on Women in Science and Engineering* (2013)
12. S. Rao, P. Rao, An overview of Hindustani music in the context of computational musicology. J. New Music Res. **43**, 1 (2014)
13. S. Chakraborty, G. Mazzola, S. Tiwari, M. Patra, *Computational Musicology in Hindustani Music* (Springer 2014)
14. F. Lerdahl, R. Jackendoff, *A Generative Theory of Tonal Music* (MIT Press, 1983)
15. H. Honig, *Musical Cognition: A Science of Listening* (Transaction Publishers, 2011)
16. J. Bor et al., The Raga Guide: A Survey of 74 Hindustani Ragas, Nimbus Records with Rotterdam Conservatory of Music (1999)
17. R. Scruton, *The Aesthetics of Music* (Oxford University Press, Oxford, 1997)
18. D.S. Rumelhart, The building blocks of cognition, in *Theoretical Issues in Reading Comprehension*, ed. by R. Spiro, B. Bruce, W. Brewer (Erlbaum Associates, Mahway, 1980)
19. R.O. Gjerdingen, *A Classic Turn of Phrase* (University of Pennysylvania Press, Philadelphia, 1988)
20. M.A. Schmuckler, Tonality and contour in melodic processing, in *The Oxford Handbook of Music Psychology* (Oxford University Press, 2016), pp. 143–152
21. W.J. Dowling, D.L. Harwood, *Music Cognition* (Academic Press Inc, Florida, 1986)
22. P. Ball, *The Music Instinct: How Music Works and Why We Can't Do Without It* (Vintage Books, London, 2011)
23. C. Seeger, Prescriptive and descriptive music-writing. Music Quart **44**(2), 184–195. https://www.jstor.org/stable/740450. Accessed 18 May 2020
24. B. Bel, Pitch Perception and Pitch Extraction in Melodic Music, ISTAR Newsletter No. 3–4, 1984–85, pp. 54–59, New Delhi. http://bakesociety.net/istar-newsletter-1984-85/
25. S. Rao, W. van der Meer, *Music in Motion: The Automated Transcription for Indian Music*, [online]. http://autrimncpa.wordpress.com/
26. W. van der Meer, *Praat manual (for musicologists)*, http://thoughts4ideas.eu/praat-manual-for-musicologists/
27. J. Salamon, E. Gómez, Melody extraction from polyphonic music signals using pitch contour characteristics. IEEE Trans. Audio Speech Lang. Process. **20**(6), 1759–1770 (2012)
28. C. Cannam, C. Landone, M. Sandler, Sonic visualiser: an open source application for viewing, analysing, and annotating music audio files, in *Proceedings of the ACM Multimedia 2010 International Conference*
29. N. Magriel, L. du Perron, *Songs of the Khayāl*, vol. 2 (Manohar Publishers and Distributors, New Delhi, 2013)
30. B. Bel, *Musical Accoustics: Beyond Levy's "Intonation of Indian Music"*, ISTAR Newsletter No. 2, (1984), pp. 7–12, New Delhi. http://bakesociety.net/istar-newsletter-1984-85/
31. W.J. Arnold, *Playing with Intonation*, ISTAR Newsletter No. 2, 1984, New Delhi (1985), pp. 61–62. http://bakesociety.net/istar-newsletter-1984-85/
32. W.J. Arnold, J. Bor, Wim v.d. Meer, On Measuring Notes: A Response to N.A. Jairazbhoy, ISTAR Newsletter No. 3–4, 1984, New Delhi (1985), pp. 46–51. http://bakesociety.net/istar-newsletter-1984-85/
33. S. Rao, *Acoustical Perspective on Raga Rasa Theory* (Munshiram Manoharlal Publishers Pvt. Ltd., New Delhi, 2000)
34. Ramashraya Jha, http://www.parrikar.org/music/marwa/jha_marwaspeak.mp3
35. R. Jha, *Abhinava Gītanjali*, vol. 5 (Sangīt Sadan Prakashan, Allahabad, 2007)
36. R.P. Parrikar, *The Marwa Matrix Part 1* https://www.parrikar.org/hindustani/marwa/
37. Vijay Koperkar http://www.shadjamadhyam.com/phrases_for_raag_puriya
38. J. Bor et al., *The Raga Guide: A Survey of 74 Hindustani Ragas* (Nimbus Records with Rotterdam Conservatory of Music, 1999), p. 114
39. J.G. Roederer, *Introduction to the Physics and Psychophysics of Music* (Springer Science and Business, 2012), p. 88

40. M.Clayton, L. Leante, Embodiment in musical performance, in *Experience and Meaning in Musical Performance* ed by M. Clayton, B. Dueck, L. Leante, (Oxford University Press, 2013)
41. https://www.parrikar.org/hindustani/bhoopali/
42. J. Bor et al., *The Raga Guide: A Survey of 74 Hindustani Ragas* (Nimbus Records with Rotterdam Conservatory of Music, 1999), p. 44.
43. I. Deliège, M. Mèlen, D. Stammers, I. Cross, Musical schemata in real-time listening to a piece of music. Music. Percept. **14**(2), 117–160 (1996)

Machine Learning Approaches to Music

Music Feature Extraction for Machine Learning

Makarand Velankar and Parag Kulkarni

Introduction

Music is fundamentally an artistic expression of communication used mostly for entertainment. It induces various emotions in the listeners and is influential in changing or elevating moods. Automatic music emotion recognition (MER) is a difficult problem, taking into account its different dimensions. These include multimodal and multifaceted aspects of music on the one hand, and listener's perception, which is subjective, on the other hand. The ground truth, i.e., that obtained by observation, involves human perception. It is influenced by diverse factors such as cultural background, likes and dislikes, current emotional state and musical background. In the task of inter-rater agreement for music similarity, the upper bound of the agreement was observed as 80% [8]. Music perception in terms of similarity or emotions is subjective as each individual will have different pre-notions built on the basis of cultural background and familiarity of the artist or music genre. Therefore, a similar upper bound is likely for emotional agreement, considering the subjectivity element associated with perceived emotions.

Each emotion has a degree or intensity associated with it. Various factors contribute to emotions that result from music. The mood is an alternative term utilized with similar meaning in the musical world. Emotions are expressed such as anger, love, happiness, sadness, romance whereas feelings can be classified as positive, negative or neutral. Sometimes the mood and emotion are used interchangeably as synonymous terms for the expression of music. Emotions themselves are a very complex and subjective phenomenon, and various approaches are explored in the

M. Velankar (✉)
Cummins College of Engineering, Pune 411052, India
e-mail: makarand.velankar@cumminscollege.in

P. Kulkarni
Kvinna Limited, Pune 411005, India

A. Salgaonkar and M. Velankar (eds.), *Computer Assisted Music and Dramatics*, Advances in Intelligent Systems and Computing 1444,
https://doi.org/10.1007/978-981-99-0887-5_4

literature to extract the emotional content of music [10, 23] Modeling emotions is a challenge, considering the large interrelated taxonomy of emotions.

In the case of popular music, lyrics do play an important role in the emotion perception and analysis of text in the lyrics is useful in emotion modeling. Lyrics, music notations used, social media comments or tags on the web or video if available for the songs in popular music are different sources of information which can be used to model the emotions and this approach is referred as multimodal analysis of music [25]. This article focuses on the role of machine learning classifiers related to emotional modeling of music with the help of acoustic features.

Over the previous decade, extensive research has been done on content analysis extracting different acoustic features from an audio signal, since music communicates feelings in a succinct but powerful manner. Individuals select music according to their choice of moods and feelings. It suggests the need to classify music in accordance with moods. Since different people have distinctive interpretations about characterizing music according to mood, it turns into a substantially more difficult task. Utilizing sound features to characterize mood can be valuable for recommendation or play-list generation.

Features play a crucial role in MER, and success depends very much on the features used in the process. Feature extraction, selection, and engineering are the three diverse approaches used. Feature extraction refers to converting high dimensional feature space to low dimensions using algorithms such as principal component analysis (PCA) or linear discriminant analysis (LDA). This new feature space includes features that are a combination of the original features. It is often hard to map new features with original ones. Thus, original features get lost in this process and analysis based on them becomes difficult in the extraction approach [22].

Feature selection methods can be categorized considering different perspectives [3]. As shown in Fig. 1, different views are on the basis of a label, search strategy, and general approach. Algorithms can be divided into supervised, unsupervised or semi-supervised categories depending on the basis of labelled or unlabelled data. It is a process of selecting relevant features from a feature space, like subset selection from the set of available features. Feature selection is a major critical step in many machine learning applications and it has a profound impact on the accuracy of machine learning classifiers. It attempts to reduce feature dimensions by removing unrelated and redundant features for a specific task. Universal methods used for feature selection are filtering, wrapper approach or embedded approach [15]. The filter approach attempts to weigh features without considering classification methods. Relief, information gain, and Fisher score constructed methods are used in the filter approach. The wrapper approach attempts to select features considering a fixed learning approach. These methods become computationally expensive if features and data are of large size. The embedded approach attempts to overcome this problem and integrates feature selection as a part of the training.

The feature selection approach chooses a subgroup of related features from the available features by using methods such as ranking or power of discrimination. These approaches attempt to reduce the computational cost and improve the accuracy of classification. Feature engineering aims at designing new features with the help of

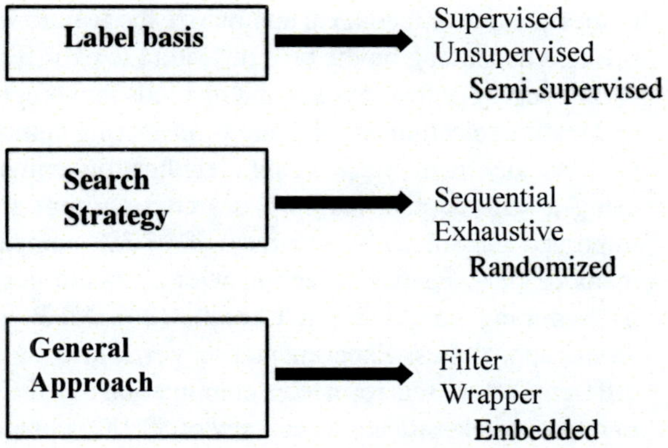

Fig. 1 General feature selection methods

domain experts to perform a specific task. Different feature ranking methods and feature selection techniques are explored in machine learning. "The only way to be sure that the highest accuracy is obtained in practical problems is testing a given classifier on a number of feature subsets, obtained from different ranking indices" [15]. A machine learning model is trained for the classification to achieve maximum accuracy, so the choice of a suitable model becomes another important parameter.

This paper attempts to validate the hypothesis that choice of features and machine learning model influences the accuracy of the MER classification significantly and the experiments are performed to find the best approach for feature selection and suitable machine learning classifier for MER application. The popular emotional classes were identified from various internet sources and music streaming apps for Indian Hindi film songs. It was observed that happy, sad, romantic, exciting and devotional are the most commonly used emotional tags in Indian film songs. A dataset of 100 songs per class is prepared with a total of 500 songs, each of about 4–5 min duration. Acoustic features are extracted using the JAudio toolbox and for training and testing, WEKA software is used with supervised learning. The hypothesis is tested for different feature sets and machine learning models using a tenfold cross-validation approach. Feature engineering is used to create novel features based on intensity and melodic information. These novel features were added in the feature sets to test improvement in the accuracy of the machine learning classifiers.

Literature Survey

Feature selection for audio music data is an extremely important step for automatic music classification or identification. Using a digital signal processing approach, features can be extracted in the time domain or frequency domain. Features can be either localized as frame-level (a small segment of audio) or globalized for the entire audio file. Global features are computed using aggregation methods of statistics such as mean, mode or median to summarize the feature values. Aggregation methods are selected depending on the type of feature to extract pertinent global information.

In a survey of feature selection methods for computational music, it was proposed that feature selection be performed considering monophonic and polyphonic features [18]. An automatic music genre classification with a local feature selection method using a self-adaptive harmony search algorithm and wrapper approach provided improved classification accuracy of 97.2% compared to other feature selection methods [11]. Feature selection with a feature engineering approach for a query by humming provided better results [24]. MER is an active research area with varied approaches. Experimentation performed on MoodSwings Lite corpus with different feature sets revealed a combination of MFCC and spectral contrast resulted in improved classification accuracy of 50.18% compared to individual features [21]. Multiple feature domains such as audio, lyrics, audio, and tags are explored for feature selection and the proposed combination of them to enhance results for MER [14].

Novel audio features related to performance expression for performance improvement, in addition to standard audio features were proposed with available frameworks with 8 categories, and the top result was obtained with 29 novel features and 71 baseline features with SVM (Support Vector Machine) classifier as the only classifier tested during the experiment [16]. A smoothed rankSVM method was proposed based on valence arousal emotion model to predict the ranking of music [6] for a dataset of 100 experimental and 40 genre music clips. Experimentation on Hindi music collections for mood analysis [17] achieved an accuracy of 48% for five mood classes considered for 250 songs. Similar experiments and results produced an accuracy of 57.1% for 4 mood classes using lyrics information [27]. Similar experiments are carried out by researchers in MER [12, 20].

To classify EEG data related to explicit moods, different machine learning classifiers such as KNN (K-nearest neighbours), regression tree, Bayesian network, SVM and ANN (artificial neural networks) were tested for accuracy. It was observed that SVM provided the highest accuracy. "Different machine learning techniques had different accuracies, and this can be due to differences of the respective algorithms used" [19]. During similar work with a comparison of decision tree J48 classifier and KNN for emotional states such as 'Sad', 'Dislike', 'Joy', 'Stress', 'Normal', 'No-Idea', 'Positive' and 'Negative', J48 classifier performance was found better than that for KNN [13]. Similar experiments are done for diverse machine learning tasks such as prediction and detection in domains other than computational music, for comparison of machine learning classifiers. Random forest algorithms provided the best result overall, but it was observed that none of the classifiers provided the best performance across different datasets [5].

From the literature survey, it was observed that for different machine learning tasks, the role of feature selection along with a choice of machine learning classifier is important. This has provided the basis to form and test our hypothesis for MER with different feature sets and classifiers. For the dataset, we used several music streaming apps and websites to obtain Hindi film songs. The test dataset is generated from the collection on the website hindigeetmala.com, and the classification of emotions is happy, sad, romantic and devotional (prayer), with a sufficient number of songs in each category. Exciting or danceable songs are downloaded from the websites baztro.com and giribalajoshi.blogspot.com. It is presumed that the song classification provided is correct and classes are generated accordingly. Extracting and selecting feature sets for the experiments was the next step after dataset generation.

Feature Selection

Audio features are extracted using the Jaudio tool for experimentation. These features are overall values calculated for the entire audio file of a song. Frame level feature values are not considered and only summary features are used. Jaudio is used for feature extraction by several researchers in music classification tasks [2, 9, 26]. Detailed information about the features supported for each tool is available in the documentation provided on the web. In our experiment, different feature selection approaches were used to select features from the set of a total of 72 feature values obtained using Jaudio. These are referred to as all features (AF).

The filter approach is used during the initial experimentation here by assigning weights to the features and ranking approach. Various algorithms for selection of features are used in the filter approach. These algorithms assign weights to the features on the basis of discrimination for different classes and similarity association for the same classes. Different ranking algorithms based on feature weights are further applied to order the features considering their discriminating power for the given dataset.

Three feature sets are selected from the superset with different approaches as mentioned in the following sub-sections. In addition to these features, 11 features based on melody and intensity were extracted to test further improvements in accuracy, if any.

Distinct Feature Set (DFS)

Similar features identified from the set of features, and features with similar intended meaning with a minor variation, are grouped together. A total of 27 distinct features are identified from the feature superset used to select the features [4]. These features are zero-crossing, root mean square, fraction of low amplitude frames, spectral flux, spectral roll-off, compactness, method of moments, 2D method of moments, Mel frequency cepstral coefficients and beat histogram features. The assumption used to select the features is that all distinct features are likely to contribute to emotional classification.

Relevant Feature Set (RFS)

Relevant features are identified considering the feature appropriateness for the emotion classification task. Features appropriateness depends on the task in hand. A set of features will vary for various tasks such as instrument or singer identification, style identification, genre classification, and emotional classifications. A set of 17 relevant features are identified from the superset with help of music emotion recognition (MER) survey [28]. This approach is suitable if the set of features is manageable with less than 100 features in general. In case the number of features is very large, such as more than 1000, this approach is not suitable.

Subset Evaluator Feature Set (SEFS)

The subset evaluator approach is used to evaluate the value of a subset of attributes by considering the specific analytical capability of each feature along with the degree of redundancy among them [29]. Subgroups of features that are closely associated with the class and having a small intercorrelation are preferred. Close association within the class refers to a similar range of values within the class with less overlap for the range values of other classes. This approach attempts to ensure better classification accuracy by selecting attributes with more discriminating capacity. This feature selection approach provided 28 features from the superset.

Machine Learning Models

Many machine learning classifier models are available, and different classifiers have their own distinctive characteristics. For this experiment, six classifiers are selected considering their usage for various applications in music classification. Since the objective here is to analyze the effect of feature selection, popular classifiers are used for the experiments.

Naive Bayes (NB), multilayer perceptron (MLP), support vector machine (SVM), J-48 decision tree (J48) and random forest (RF) classifiers are used in many experiments and are used for the experimentation in this work. Fine-tuning various parameters of the classifier may result in improved performance. We have retained the default parameter values for the experimentation in the WEKA tool[1]. A detailed description of machine learning classifiers is available in a number of textbooks and on web resources [1, 7].

A tenfold cross-validation methodology is used to evaluate the accuracy of each classifier for a specific feature set. Different parameters are used for determining the accuracy of a classifier. Correctly classified instances from the dataset are one

[1] Popular Machine learning tool.

Table 1 Accuracy measures for MLP classifier for AF set

Sr. No	Accuracy measure	Value
1	Correctly classified instances	301
2	Accuracy in percentage	60.2%
3	Mean absolute error	0.1636
4	Root mean squared error	0.3611
5	Precision	0.605
6	Recall	0.602
7	F-measure	0.602
8	ROC Area	0.859

of the simplest measures which provide accuracy. Classification accuracy is one of the important measures for comparison of performance, expressed as percentage. An example of different accuracy measures for MLP classifier results with all features (AF) selected is as shown in Table 1. It shows different measures used for the performance evaluation of the classifiers used in machine learning. Classification errors are basically the difference between the predicted and the actual values during the classification. Mean absolute error provides an average of all absolute errors and root mean squared error represents the standard deviation of the difference in predicted and actual values. Precision, recall and F-measure are common evaluation measures in various domains such as machine learning, information retrieval, pattern recognition etc. Precision refers to the fraction of correct predictions in the actual predictions and recall refers to the fraction of correct predictions from the actual correct data items. F-measure represents a single value as a combination of precision and recall using harmonic mean. ROC (Receiver Operating Characteristic) area is a plot of sensitivity (true positive rate) vs specificity (false positive rate) with values ranging from 0 to 1. These different measures represent the classifier performance for the given labelled dataset in machine learning.

Confusion matrix (error matrix) is an important performance measure which helps to understand a difference between the actual classification and predicted classification for each class element which can further be used to find true positives, false positives, true negatives and false negatives. Class-wise accuracy can be observed using a confusion matrix.

Table 2 provides a confusion matrix for MLP as a classifier with all features (AF). The main diagonal of the matrix shows correct predictions as 31, 88, 67, 52 and 63 for the classes Romantic (R), Exciting (E), Happy (H), Sad (S) and Devotional (D), respectively. Accuracy is calculated for a total of 500 predictions, out of which 301 are correct predictions. The ideal confusion matrix with 100% accuracy is a diagonal matrix with only diagonal elements for a principal diagonal and all other elements as 0. Similar results are obtained for each classifier with different feature sets. Classifier accuracy in percentage is used as the main comparison parameter for the result evaluation for feature selection.

Table 2 Confusion matrix for MLP classifier for all feature (AF)

Predicted Class → / Actual Class ↓	R	E	H	S	D
R	31	4	3	38	24
E	3	88	6	3	0
H	4	13	67	15	1
S	31	2	8	52	7
D	24	2	2	9	63

Results and Discussions

Experiments were performed for four different feature sets: superset (AF) with all 72 features, distinct feature set (DFS), relevant feature set (RFS) and feature set from a subset evaluator (SEFS), as explained in the previous section. The results are evaluated for five machine learning classifiers, as shown in Table 3.

It can be noticed from Table 3 that the choice of classifiers and the feature sets has a significant impact on the classification accuracy, which varies from 51.2 to 70%. This supports the hypothesis about the significant impact on the performance accuracy with 18.8% variation considering the minimum and maximum accuracy obtained. The average accuracy for each feature set indicates the performance of the feature set for different classifiers. DFS feature set has provided a maximum average accuracy with 61.75%. Individual classifier average performance is varied. MLP provided maximum accuracy followed by SVM. A combination of MLP and DFS provided a maximum accuracy of 70%.

From the confusion matrix, the major conflict was observed in Sad and Romantic classes. Careful observation and analysis of the dataset by domain experts revealed

Table 3 Comparison for different feature sets and machine learning classifiers

Machine learning classifier	Feature Set (FS) accuracy				
	AF	DFS	RFS	SEFS	
Naive bayes	54.4	51.2	48.4	53.8	52.2
MLP	60.2	70	64	63.8	64.5
SVM	64	62.6	59.2	62	61.95
J-48	53.2	53.6	49.2	54.8	52.7
Random forest	61.2	60.8	57.6	60.6	60.05
Average accuracy	58.6	61.75	55.68	59	

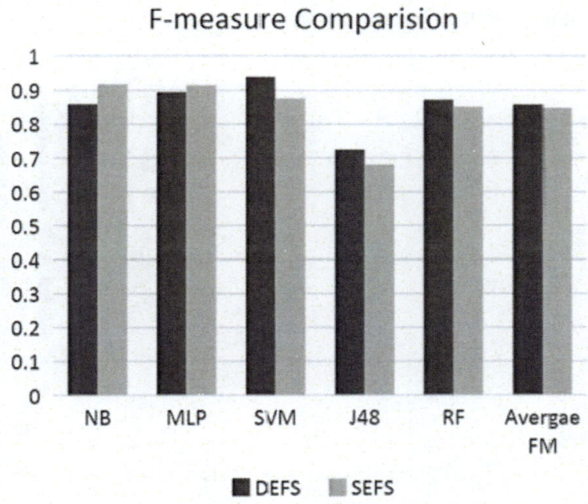

Fig. 2 F measure comparison

that many songs belonged to both classes and therefore there is a major overlap. A possible solution could be generating a new class as a combination of both emotions, as a Sad Romantic class, in addition to individual classes. Another major conflict was between the Devotional and Romantic classes and hence there is a need for some additional features to reduce conflict. Features from the lyrics can be added to improve accuracy in such cases.

After observing the accuracy for different feature sets, a further comparison is made for DFS and SEFS, for other evaluation measures. F measure indicates the combination of precision and recall with harmonic mean of both. Figure 2 shows a comparative analysis of F measure for different classifiers with the feature sets DFS and SEFS. The average F measure is higher for DFS compared to SEFS, as can be observed from Fig. 2.

Similarly, ROC or area under the ROC curve is another measure of performance of classifiers with a range from 0 to 1. The more the value of ROC, i.e. close to 1, the better is the classifying accuracy of the classifier. Figure 3 shows the comparative performance of feature sets DFS and SEFS for various classifiers. It can be observed that SEFS has a marginally better average ROC compared to DFS. The DFS feature set has an average F measure value of 0.858 which is marginally higher compared to 0.8484 for the SEFS feature set; whereas the SEFS feature set has a better average ROC value, at 0.9408, compared to 0.9396 for DFS.

This indicates that comparison of different parameters may reveal diverse results and it depends on the importance of measures for the task to be done. The results indicate that more experimentation with a different feature set may provide better accuracy for classification. Similar experimentation with various datasets and features can be useful for supporting the findings and observations discussed for this pilot experiment.

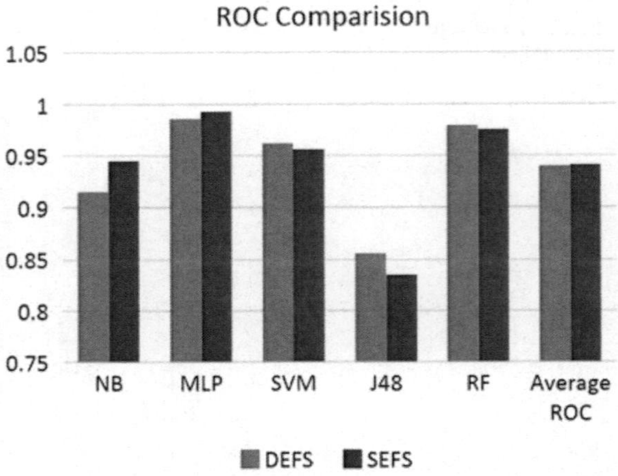

Fig. 3 ROC comparison

Conclusion and Future Directions

For the classifiers we considered, the accuracy obtained is different for various feature sets. Deciding an appropriate classifier and feature set is an interesting task and what should be the exact number of features or which the feature selection method is best has no definite answer. It depends on many parameters such as the task in hand, size of the data set, types of features, etc. The present experiments revealed that a feature selection with the help of DFS with additional features as melody and amplitude has provided 73.6% as the maximum accuracy for MER with 5 classes. The DFS approach with additional descriptive features has an edge over the feature selection methods compared by us. MLP machine learning classifier based on a neural network performed better with an overall maximum accuracy of 64.5%.

Feature selection with the help of distinct features may not be a suitable approach when the number of features is large. The inputs from domain experts for feature selection are certainly useful for improving the performance or to fine-tune the system. Feature selection algorithms suggest feature set selection using some evaluation techniques to rank the features and experiments with various algorithms using a filter, wrapper or embedded approaches may lead to improved performance.

Features are truly the heart of any machine learning classification system. The better the features; the better will be the performance of a system. Feature engineering with the help of domain experts to design new features is a useful approach to improve performance. The time invested in a feature selection or reduction before the selection of a classifier is really worthwhile and likely to provide improved results.

The choice of machine learning classifier is a tricky decision and needs experimentation before any conclusion. From our experiments, it is revealed that MLP and SVM are better choices among the 5 classifiers considered for MER tasks. We propose further experiments with the addition of a new class, a combination of sad and romantic emotions as a possible solution to reduce overlap and improve accuracy. Novel features based on a specific genre can be generated and experimented further to improve the accuracy for MER.

References

1. E. Alpaydin, *Introduction to Machine Learning* (MIT press, 2014)
2. J. Barbosa, C. McKay, I. Fujinaga, Evaluating automated classification techniques for folk music genres from the Brazilian northeast. Comput. Music: Beyond Front. Signal process. Comput. Models (2015)
3. G. Chandrashekar, F. Sahin, A survey on feature selection methods. ACM J. Comput. Electr. Eng. 16–28 (2014)
4. D. McEnnis, C. McKay, I. Fujinaga, P. Depalle, Jaudio: a feature extraction library, in *ISMIR* (2005)
5. T. M. Deist, F.J.W.M. Dankers, G. Valdes, R. Wijsman, C. Hsu, C. Oberije, T. Lustberg et al.,Machine learning algorithms for outcome prediction in (chemo) radiotherapy: an empirical comparison of classifiers. Med. Phys. **45**(7), 3449–3459 (2018)
6. J. Fan, K. Tatar, M. Thorogood, P. Pasquier, Ranking-based emotion recognition for experimental music, in *ISMIR* (2017), pp. 368–375
7. P. Flach, *Machine Learning: The Art and Science of Algorithms that Make Sense of Data.* (Cambridge University Press, 2012)
8. A. Flexer, T. Grill, The problem of limited inter-rater agreement in modelling music similarity. J. New Music Res. **45**(3), 239–251 (2016)
9. J. Grekow, Emotion detection using feature extraction tools, in *International Symposium on Methodologies for Intelligent Systems* (Springer, Cham, 2015), pp. 267–272
10. J.-L. Hsu, Y.-L. Zhen, T.-C. Lin, Y.-S. Chiu, Affective content analysis of music emotion through EEG.Multimed. Syst. **24**(2), 195–210 (2018). http://giribalajoshi.blogspot.com/2014/12/top-bollywood-dance-numbers-of-2014.html?m=1. Accessed 16 Feb 2019 for dataset; https://www.baztro.com/top-70-best-bollywood-dance-party-songs-list-latest-2016/. Accessed on 16 Feb 2019 for dataset; https://www.hindigeetmala.net/category/. Accessed on 16 Feb 2019 for dataset
11. Y.F. Huang, S.M. Lin, H.Y. Wu, Y.S. Li, Music genre classification based on local feature selection using a self-adaptive harmony search algorithm. Data Knowl. Eng.
12. J.H. Juthi, A. Gomes, T. Bhuiyan, I. Mahmud, Music emotion recognition with the extraction of audio features using machine learning approaches, in *Proceedings of ICETIT 2019* (Springer, Cham, 2020), pp. 318–329
13. A.M. Khan, M. Lawo, From physiological data to emotional states: conducting a user study and comparing machine learning classifiers. Sens. Transducers **6** (2016)
14. Y.E. Kim, E.M. Schmidt, R. Migneco, B.G. Morton, P. Richardson, J. Scott, J.A. Speck, D. Turnbull, Music emotion recognition: a state of the art review, in *Proceedings of ISMIR* (2010), pp. 255–266
15. J. Novaković (2016) Toward optimal feature selection using ranking methods and classification algorithms. Yugoslav J. Oper. Res.
16. R. Panda, R.M. Malheiro, R.P. Paiva (2018) Novel audio features for music emotion recognition. IEEE Trans. Affect. Comput.
17. B.G. Patra, D. Das, S. Bandyopadhyay, Unsupervised approach to Hindi music mood classification, in *Mining Intelligence and Knowledge Exploration* (Springer, Cham, 2013), pp. 62–69
18. J. Pickens, A survey of feature selection techniques for music information retrieval, in *Proceedings of the 2nd International Symposium on Music Information Retrieval (ISMIR)* (2001)

19. S. Qureshi, J. Hagelbäck, S.M.Z. Iqbal, H. Javaid, C.A. Lindley, Evaluation of classifiers for emotion detection while performing physical and visual tasks: tower of hanoi and IAPS, in *Proceedings of SAI Intelligent Systems Conference* (Springer, Cham, 2018), pp. 347–363
20. R. Sarkar, S. Choudhury, S. Dutta, A. Roy, S.K. Saha, Recognition of emotion in music based on deep convolutional neural network. Multimed. Tools Appl. 1–19 (2019)
21. E.M. Schmidt, D. Turnbull, Y.E. Kim, Feature selection for content-based, time-varying musical emotion regression, in *Proceedings of the International Conference on Multimedia Information Retrieval* (ACM, 2010), pp. 267–274
22. J. Tang, S. Alelyani, H. Liu, Feature selection for classification: a review, in *Data Classification: Algorithms and Applications* (CRC Press, 2014), pp. 37–64. https://doi.org/10.1201/b17320
23. N. Thammasan, K. Moriyama, K.-I. Fukui, M. Numao, Continuous music-emotion recognition based on electroencephalogram. IEICE Trans. Inf. Syst. **99**(4), 1234–1241 (2016)
24. M. Velankar, P. Kulkarni, Feature engineering and generation for music audio data. IJETT **5**(1) (2018)
25. V.K. Velankar, P. Kulkarni, Universal music context representation with multi-modal analysis for efficient retrieval using parallel processing paradigm, in *Proceedings of DLFM* (2019)
26. M. Weber, T. Krismayer, J. Wöß, L. Aigmüller, P. Birnzain, JKU-Tinnitus approach to emotion in music task, in *MediaEval* (2015)
27. X. Yang, Y. Dong, J. Li, Review of data features-based music emotion recognition methods. Multimed. Syst. **24**(4), 365–389 (2018)
28. Y.H. Yang, Y.C. Lin, H.T. Cheng, I.B. Liao, Y.C. Ho, H.H. Chen, Toward multi-modal music emotion classification, in *Pacific-Rim Conference on Multimedia* (Springer, Berlin, Heidelberg, 2008), pp. 70–79
29. P. Yildirim, Filter based feature selection methods for prediction of risks in hepatitis disease. Int. J. Mach. Learn. Comput. **5**(4), 258 (2015)

Role of Prosody in Music Meaning

Hari Sahasrabuddhe

Background

Communication between humans occurs at multiple levels, such as visual, vocal, and verbal. Visual aspects refer to parameters like body language and facial expression. Vocal represents qualities such as the tonal info, stress, and speed of delivery. Verbal refers to actual words or sentences i.e., the content of the talk. Nonverbal signals also play a significant role in the recipient's interpretation of a message [1]. In the case of speech, the intended meaning is conveyed through the sentences with some nonverbal cues. These nonverbal cues are broadly referred to as prosody [2]. The impact of nonverbal components such as vocal or tonal cues and visual or body language is reported in various research articles [3–5]. Prosody in linguistics refers to the elements of speech that are not individual phonetic segments, i.e., vowels and consonants but are properties of syllables and larger units of speech such as intonation, tone, stress, and rhythm [Wikipedia]. Speech prosody in association with 12 different emotions is explored with 2 different cultural perspectives [6]. The rhythm along with intonation holds prosodic information. For example, rising and falling intonation contours signal the speaker's intended meanings [7]. Arguably, a monotonous speech is significantly more difficult to comprehend than speech with a usual inflections of pitch and loudness. In Indian usage, an indicative sentence can be turned into an interrogative form without changing word order, simply by raising the pitch towards the end of the utterance, as in "You are going to a movie?".

Prosody indicates an emotion conceived by the speaker as well as aroused in the listener. A typical political speech contains many examples of this latter phenomenon. In prosody there may signals as to how confident a speaker is or how strongly a statement is asserted or is it a sarcasm or the speaker does not really mean it. Often

H. Sahasrabuddhe (✉)
Former Professor IITB, Powai, Mumbai, Maharashtra, India
e-mail: hvs_buddhe@hotmail.com

A. Salgaonkar and M. Velankar (eds.), *Computer Assisted Music and Dramatics*, Advances in Intelligent Systems and Computing 1444,
https://doi.org/10.1007/978-981-99-0887-5_5

one word is representative of the meaning of a statement. However, this keyword may occur anywhere in the utterance according to the dictates of the correct form. The task of casting it as the central theme is a part of prosody. Relationships between parts of a sentence can also be constructed with the help of prosody. For example, without prosody how meaningful would be Hamlet's famous line, "Look here upon this picture, and on this."?

The informal role of prosody in prose is, to a large extent, has been taken over by meter in verse. Change of meter could signal a shift in the topic, a change in the situation or in the pace of action. Readers can probably find several examples of this phenomena in the poetry they know.

"Meaning" as it applies to music is a whole topic by itself. Does there exists a meaning in it, could be a question regarding the more abstract forms of music? Yet, the music of Beethoven's 9th symphony must have had a powerful meaning for himself that he conveyed by breaking the convention and including words in it along with the instrumental pieces. Music develops its affective meaning for the listener through musical structures that are in some sense homologous to structures of interpersonal behaviour. So, at one level of abstraction, music and interaction are similar in their underlying patterns of emotional dimensionality [8]. The role of ethnomusicology and interpretation of music with styles and social context has been explored [9]. A study of Jazz performances has revealed variations, dynamics and aesthetics in musical interactions [10]. The study of aesthetic elements in speech and music prosody using a holistic approach helped in understanding amusia, the in-ability to identify and produce musical tones [19].

Like speech, music performance has different auditory characteristics that convey meaning. Music is traditionally being used to convey a message with emotional appeal in the form of auditory rendition. Prosody in music is attributed to different facets of music excluding the lyrics [11]. These different facets such as melody, rhythm, intonation and music ornamentations contribute to conveying the intended meaning. Psychoacoustic prosody features such as loudness, tempo, melody contour, spectral centroid, spectral flux, sharpness, roughness and their strong association with emotions are explored for speech and music [12]. While concluding this Section, I quote from [13], "Musical prosody is a complex, rule-governed form of auditory stimulation that can move listeners emotionally in systematic ways".

In the next section let us discuss three parameters that contribute to the prosody in music. The third section is on discussion and suggestion and the fourth is the conclusion of the paper.

Parameters of Music Contributing to Prosody

Tempo and Rhythm

A quicker tempo generally signifies an excitement or happy mood; a slower one peace and quiet, sadness or caution. The tempo is measured in beats per minute (BPM) and is associated with the rhythm. Various techniques such as autocorrelation and novelty curve for onset detection, real time beat tracking, etc., are used to estimate the tempo [14, 15].

The tempo associated with rhythm provides meaning to the rhythmic patterns in music. Children at play often repeat a 3-semitone drop (e.g., E-C) to tease; sirens of emergency vehicles employ a 3-semitone interval. So do Basketball spectators, when they shout "airball". Arguably, this is not due to a mere coincidence. The time signature is a convention used in western notation to specify the role of meter used along with beats. Simple, compound or complex rhythms can be represented using different conventions. The most common simple time signatures are 4/4, 2/4 and 3/4, while 5/4, 7/4 and 7/8 are the complex ones. The repetitive structures with an emphasis on the specific pattern in music conveys a specific meaning.

Instrumentation

There exists a mapping between the sound of a specific instrument and emotions. The Shehnai in the north and Naadswaram in south India are played on the auspicious occasions. They form the important aspect of temple service. It is but natural that the introduction of a short passage or even a single note on an instrument like shehnai or naadaswaram will steer an Indian mind towards the positive thoughts. The readers experienced of Christianity may find churchbells leading their thoughts in fixed directions.

Besides this association, sounds of instruments have different characters. For example, Geet Ramayana [16], a suite of 56 songs that tells the story of the epic Ramayana has been written by G D Madgulkar in Marathi; Sudhir Phadke has composed music to these songs. In a song, the dying king Dasharatha paints a picture of his total hopelessness. It is a dark song. In one stanza he remembers Rama and imagines the change in mood of the people who will meet him on his return. The change in the instrumentation in that stanza vis a vis the rest of the song is worth noting. Similar examples can be observed in different music genres with effective use of instrumentation for conveying the musical meaning.

Melody

Emotions can rise and fall with the notes of a melody. Some phrases sound happy, others sad. Melody is the heart of music and extracting melodic features is essential for a variety of applications such as similarity identification, copyright infringements, query by humming, content-based retrieval, etc. Pitch estimation is important in identifying melodic contour. Melodic contour symbolizes changes in perceived pitch over a definite time frame. Fundamental pitch estimation, overtones, just noticeable difference (JND), silence, shimmer, jitter, etc., are the parameters that have been employed in various pitch detection algorithms (PDA). The data is treated either in frequency or in time domain [17].

Meaning itself has various dimensions and is a complex phenomenon. In addition to words, all sensory inputs such as images, smells, sound can contribute to a general meaning. Douglas Hofstadter explores the question "How does ego arise in a brain" in his book, "I Am a Strange Loop" [18]. Many of the ideas in that book apply to the meaning of meaning.

Discussion and Suggestions

The prosodic information communicated through melody, instrumentation and tempo has been a part of our daily experience almost since our childhood. There will hardly be any differences of opinions about the meanings communicated by them. Maybe because of that it has not attracted researchers to critically evaluate their individual contributions. However, their application has always been based upon the intuition of the humans who employed them. With the advent of software technology, audio signal processing libraries in particular, it is possible to measure the tempo of a given piece of music very easily. The melody contours of a musical piece could suitably be interpreted in order to extract the required information from it. Given a musical piece, separating and recognizing the instruments played in it could be a research problem. Through the literature survey, it is clear that most of the studies in this domain have been undertaken in the non-Indian context. India has a rich tradition of both classical as well as non-classical (i.e., cinema, theater and folk) music. Developing annotated databases of the musical pieces from diverse varieties could be the first step towards their analysis. Building AI based applications on them could be the next.

For example, the Geet Ramayan itself is a source of sentiment invoking music. One could take up analysis of tempo, instrumentation and melody in each of its songs, rather, stanzas, and validate the contentions in this article. The findings of such experiments could be useful in the design of suitable music. One can appreciate an entity by knowing more details about the same. It could be a way to reach the meaning of the meaning. May or may not be for any material purpose, but certainly for the eternal joy of academics one should try such things out.

Conclusion

In summary, this article has introduced the concept of prosody to speech and its extension to music. A brief literature review has been presented in order to introduce the variety of work people have undertaken and its application. Tempo, instrumentation and melody in particular have been proposed as the three prosodic parameters for music. Suggestion to validate this contention in the context of Indian music in particular has been made. The examples of prosodic parameters in day-to-day life have been presented. They may inspire the readers to explore more in this domain that should enhance their music appreciation at least.

We spent a stimulating and delightful couple of days at the Vidyanagari campus of University of Mumbai in February 2019 discussing the many ways music and technology can and do interact. It is most rewarding to see the present volume emerge as an outcome of that workshop. Much time was lost, mainly because the world passed through a new type of freeze soon after that. But now that freeze finally seems to be over, and we can proceed as if nothing happened.

The reach of computing continues to expand at a breathtaking pace. Analytical studies now examine audio recordings of performance rather than be satisfied with notation, with all the limitations of the latter. This power has been applied towards varied goals from information retrieval to understanding raga form to vocology. There is even an examination of the role of computing in the production of music, an angle often overlooked. “Sangeeta”, the Sanskrit term, broadly translated as “music”, nevertheless includes dance. And the reader will find that this volume remains true to that inclusion.

I have known the editors, Ambuja Salgaonkar and Makarand Velankar, for decades now. I congratulate them on completing this notable volume and wish them a wide readership.

References

1. D. Phutela, The importance of non-verbal communication. IUP J. Soft Ski. **9**(4), 43 (2015)
2. R.L. Mitchell, E.D. Ross, Attitudinal prosody: What we know and directions for future study. Neurosci. Biobehav. Rev. **37**(3), 471–479 (2013)
3. T.W. Liew, S.M. Tan, H. Ismail, Exploring the effects of a non-interactive talking avatar on social presence, credibility, trust, and patronage intention in an e-commerce website. HCIS **7**(1), 1–21 (2017)
4. A. De Waele, A.S. Claeys, V. Cauberghe, G. Fannes, Spokespersons' nonverbal behavior in times of crisis: The relative importance of visual and vocal cues. J. Nonverbal Behav. **42**(4), 441–460 (2018)
5. N. Jackob, T. Roessing, T. Petersen, The effects of verbal and nonverbal elements in persuasive communication: Findings from two multi-method experiments (2011)
6. A.S. Cowen, P. Laukka, H.A. Elfenbein, R. Liu, D. Keltner, The primacy of categories in the recognition of 12 emotions in speech prosody across two cultures. Nat. Hum. Behav. **3**(4), 369–382 (2019)

7. A. Buxó-Lugo, D.G. Watson, Evidence for the influence of syntax on prosodic parsing. J. Mem. Lang. **90**, 1–13 (2016)
8. C.L. Ridgeway, J.M. Roberts, Urban popular music and interaction: A semantic relationship. Ethnomusicology, 233–251 (1976)
9. M. Clayton, The social and personal functions of music in cross-cultural perspective. The Oxford handbook of music psychology, 35–44 (2009)
10. M. Doffman, Making it groove! Entrainment, participation and discrepancy in the'conversation'of a jazz trio. Language & History **52**(1), 130–147 (2009)
11. S. Brown, A joint prosodic origin of language and music. Front. Psychol. **8**, 1894 (2017)
12. E. Coutinho, N. Dibben, Psychoacoustic cues to emotion in speech prosody and music. Cogn. Emot. **27**(4), 658–684 (2013)
13. C. Palmer, S. Hutchins, What is musical prosody? Psychol. Learn. Motiv. **46**, 245–278 (2006)
14. P. Grosche, M. Müller, Tempogram toolbox: Matlab implementations for tempo and pulse analysis of music recordings, in *Proceedings of the 12th International Conference on Music Information Retrieval (ISMIR)*, Miami, FL, USA (2011, October) pp. 24–28.
15. J.L. Oliveira, M.E. Davies, F. Gouyon, L.P. Reis, Beat tracking for multiple applications: A multi-agent system architecture with state recovery. IEEE Trans. Audio Speech Lang. Process. **20**(10), 2696–2706 (2012)
16. Gadima (1955) https://www.gadima.com/category/3/0/0/geetramayan-sudhir-phadke
17. V. Rao, P. Rao, Vocal melody extraction in the presence of pitched accompaniment in polyphonic music. IEEE Trans. Audio Speech Lang. Process. **18**(8), 2145–2154 (2010)
18. Douglas Hofstadter (2007) https://www.audible.in/pd/I-Am-a-Strange-Loop-Audiobook/B07HJB17HB
19. A. Lautrari, M. Lorch, *Preserved appreciation of aesthetic elements of speech and music prosody in an amusic individual: A holistic approach* (Elsevier, Brain and Cognition, 2017)

Estimation of Prosody in Music: A Case Study of Geet Ramayana

Ambuja Salgaonkar and Makarand Velankar

Background

Prosody or intonation is experienced in music as well as in speech. It is employed to convey emotions. Prosody in speech is due to the deliberate emphasis on the words and the pauses or delays that are intended to make a speech effectively. In music, prosody is the mapping of syllables to notes in the melody to which the text is sung. It is the music concerning the ambience of the lyrics and its connotation [39]. There is a strong association between speech prosody [3, 7, 16, 32] and music prosody [21, 30] Understanding prosody helps to enhance the emotional quotient. Music prosody in particular has helped rehabilitation of the persons deprived of speech and hearing [17, 19, 49].

The knowledge of prosody has been a requirement of various jobs: creating mood-wise playlists of audios, organizing prosodic cues in performance to induce emotions in the listeners and music therapy to name a few [1]. A piece of music that consists of a specific type of prosody is a need when the music is employed for therapeutic purposes [6]. Machine learning employed for generating prosodic information from a given piece of music may assist in selecting the right musical piece given the specifications of required prosody.

The work presented in this paper is about the automatic exploration of prosodic cues from Geet Ramayana [14, 15], a set of 56 songs in Marathi, depicting the life of a heroic figure in Indian mythology. During the music performance or recording of any

A. Salgaonkar (✉)
Department of Computer Science, University of Mumbai, Mumbai 400098, India
e-mail: ambujas@udcs.mu.ac.in

M. Velankar
Cummins College of Engineering, Pune 411052, India
e-mail: makarand.velankar@cumminscollege.in

A. Salgaonkar and M. Velankar (eds.), *Computer Assisted Music and Dramatics*,
Advances in Intelligent Systems and Computing 1444,
https://doi.org/10.1007/978-981-99-0887-5_6

album, the arrangement and use of instruments, improvisations, music ornamentation, variations in different musical aspects such as tempo, intensity, timbre, etc., lead to various perceptions and interpretations [5, 11, 35, 41]. Tempo, instrumentation and melody are prosodies in music [53]. We employ reverse engineering to estimate a set of parameters that generates these prosodic cues. A regression-based ensemble learning model that infers the mood of a given song is one of the contributions of this paper.

In the following sections first, we provide a brief survey of the research on prosody and its applications. Next, we provide methods of estimating a few prosodic parameters by employing general-purpose software. Following it, we present our case of estimating the parameters to know the tempo, instrumentation, melody and overall mood of the songs in Geet Ramayana.

Literature Survey

Prosody has been employed for the conversion of a neutral speech into an emotional speech. It has been shown that a linear modification model does not work, while the Gaussian mixture model is suitable for a small training set, a better output on a large set is possible by employing CART [47]. Besides the phrase boundaries, accent and sentence mood, pitch, loudness, speed of a speech, voice quality, duration, pause, rhythm, etc., are the prosodic attributes that play an important role in the speech-translation. However, to identify one or more attributes in combination with a correct meaning of a speech is a research problem [33]. Table 1 shows a mapping of prosodic parameters to the emotions [44].

Happiness is associated with rapid tempo, high pitch, large pitch range, and bright voice quality; sadness is associated with slow tempo, low pitch, narrow pitch range, and soft voice quality; anger is associated with fast tempo, high pitch, wide pitch range, and rising pitch contours; and fear is associated with fast tempo, high pitch, wide pitch range, large pitch variability, and varied loudness.

Word pronunciations are specifically articulated and learnt almost mechanically. They make a verbal expression pleasant to hear. An orator employing pause, emphasis and voice modulation to make a speech effective is an art. The former is referred to as intrinsic *prosody* while the latter is referred to as extrinsic *prosody* [2]. The inbuilt properties like pitch range limits, size and shape of the vocal cavity, teeth,

Table 1 Prosodic parameter mapping with emotions

	Happy	Sad	Anger	Fear
Tempo	Rapid	Slow	Fast	Fast
Pitch	High	Low	High	Variable amplitude
Pitch range	Large	Narrow	Wide	Wide
Voice	Bright	Soft	Rising	Varied loudness

etc., of a vocal apparatus play a vital role in producing effective prosodic cues. Consideration of these parameters adds to the complexity while identifying a best performing configuration for a vocal apparatus.

The importance of objective and quantitative modeling of tonal features have been emphasized for the purpose of the manifestation of information [12]. In music the variations in the frequency, amplitude, pitch and timbre of sound are employed to convey emotions [27]. Descending speech contours and notes of long duration essentially marks the phrase boundaries in both speech and music [36]. An evaluation of musically trained and untrained subjects of diverse age groups in decoding the prosody of a tonal utterance of an emotionally neutral sentence uttered with an emotion like happiness, sadness, fear or anger, shows that labeling an emotion as positive and negative is easier than labeling negative emotions as either anger or fear. The ability of a music person to interpret an emotional speech is on par or better than that of a drama person [48]. The discrimination ability of a non-musician person for differentiating between tuneful music and one with tampered melody or rhythm, and consequently recognizing changes in the utterance of a phrase has been examined [21]. It has been shown that music and speech perception are correlated through prosodic cues like time and stress [43]. Acoustic cues for emotions in music and speech are compared. It is observed that they are independent of familiarity with the language [22, 23]. Subjects with interesting music ranging from "I hate music" to "I have occasionally listened to music" to "I love music" to "I understand it", "I sing" were made to listen to a variety of popular emotional songs and a set of speeches compiled from film, drama, interviews and poem recitals. They were asked to make an indication whenever they felt a change in emotions. A neural network was trained to model their behaviour. It has been inferred that speech and music share five prosodic features, namely, loudness, tempo, speech-rate, melodic and prosodic contours, and spectral centroid and sharpness. Spectral flux is the sixth required feature for modelling music while roughness is for speech. Applications of these findings to the domain of health and well-being to linguistics have been pointed out [6]. A strong association between speech and musical prosody has been observed even in patients with traumatic brain injury [30]. A possibility of employing music training to improve the understanding of intonation by the hearing-impaired persons has been explored [17]. Introducing carefully articulated rhythmic components in music therapy is found to be effective for the rehabilitation of patients with non-fluent aphasia and delayed speech development [19, 49]. Various music therapy techniques are proposed based on a music prosody study [31].

Extraction of Prosodic Information from a Musical Piece

Tempo in music or speech refers to the speed or pace of a given piece. *Tempogram* is a representation of tempo. It could be employed for purposes like rhythm identification and bit tracking, or structure analysis and music classification. A tempogram based on

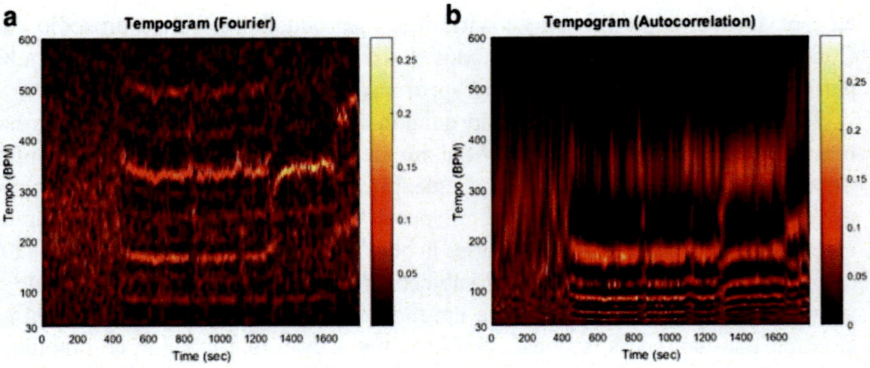

Fig. 1 Tempogram of Raga Malkauns (**a** Fourier and **b** Autocorrelation)

pursuit matching demonstrates higher resolution with stronger sparsity and flexibility [20].

Tempograms can be plotted by employing readymade libraries [18]. Figure 1a, b shows the tempograms of Raga Malkauns, an instance of Hindustani classical music (HCM).

No real tempo information can be extracted before 450 s, as can be noticed from Fig. 1a, b. This is because there is no rhythmic structure in Aalaap, the style with which a performance begins. Tempo information has been captured after the rhythm starts. Some irregularities are observed near 850, 1150 and 1300 s. An increase in tempo can be clearly observed after about 1300 s. This corresponds to the increase in energy as seen in Fig. 1b. Figure 1 helps in understanding rhythm solos that are common in HCM.

Researching variations in tempo for understanding the intended composition by a conductor and for correcting notations has been explored by researchers for a long time. Variation in tempo of the tempogram of Fig. 2 has been plotted using the INESC Porto Beat Tracker (IBT) [34], as an example.

Table 1 contains a listing of tempograms of three musical pieces that are derived from three varieties of music. The observations noted next to the tempograms are in comparison with the tempogram in Fig. 2. The noticeable consistency between the observations and the known particulars of the music is indicative of the efficacy of tempograms in music classification (Table 2).

In the absence of the instrumental rhythm in some places, the conventional onset detection algorithms fail in estimating the tempo of the overall song. We have employed the median tempo analysis (MTA) that seems to be working here satisfactorily. However, MTA may not prove to be sufficient in cases where the tempo changes frequently.

Melody is a linear sequence of notes the listener hears rhythm and pitch as a single entity. Identifying and noting down a melody accurately is a more difficult task than to remember it roughly. Generally, the predominant fundamental frequency (F0) is estimated first, and then the sequence of frequency values is normalized. Methods

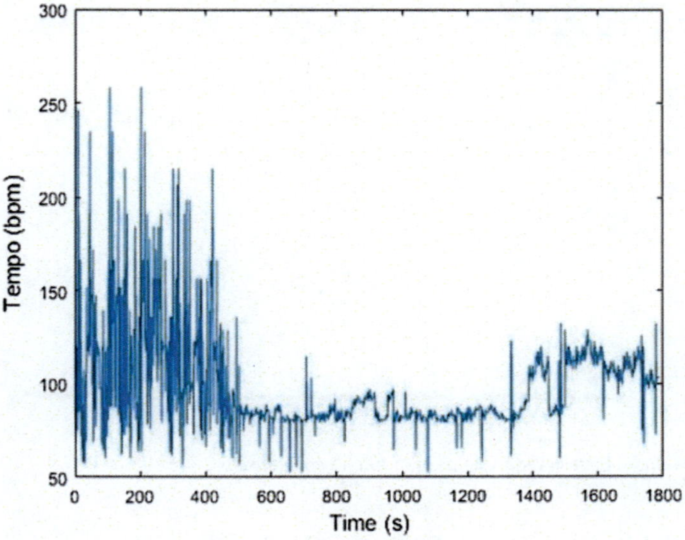

Fig. 2 Variation in tempo in Raga Malkauns using IBT algorithm

Table 2 Example of tempogram analysis for classification of music

Song ID	Particulars of the music	Tempograms	Observations about tempo
Song 1	Progressive rock, minor variations in tempo	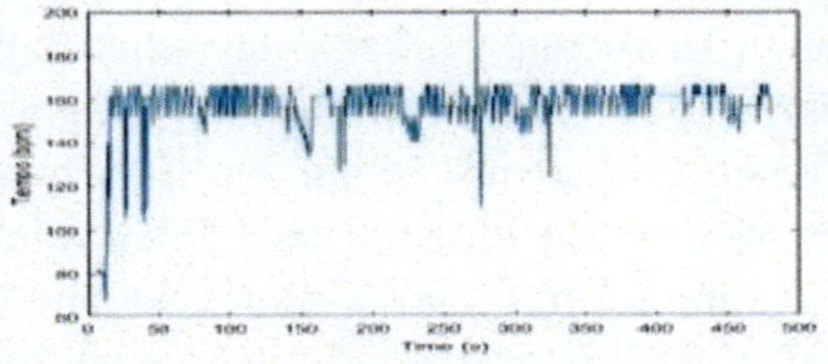	161 BPM, less variations compared to HCM
Song 2	Mainstream popular Hindi song, a very few tempo variations	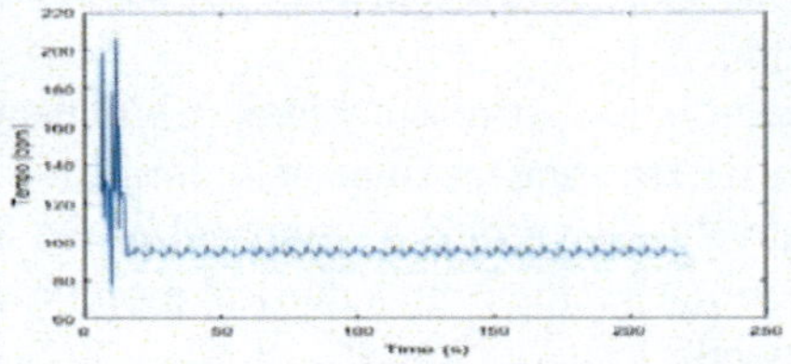	96 BPM, comparable with the HCM case, variations is less
Song 3	Mainstream Indie rock, structure similar to HCM, instrumental rhythm is introduced late, constant tempo	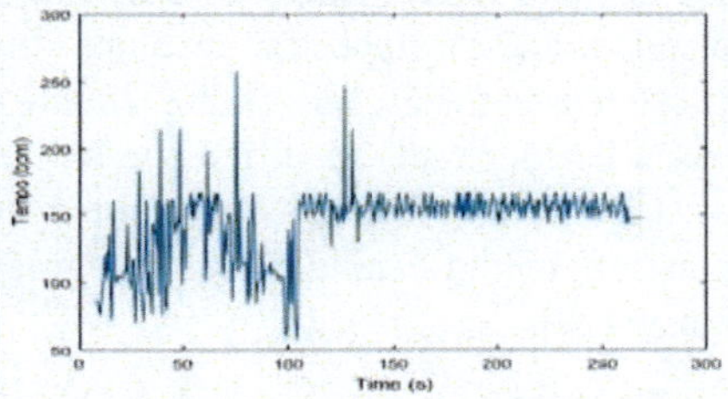	152 BPM, difficult to compute till the first 100 s

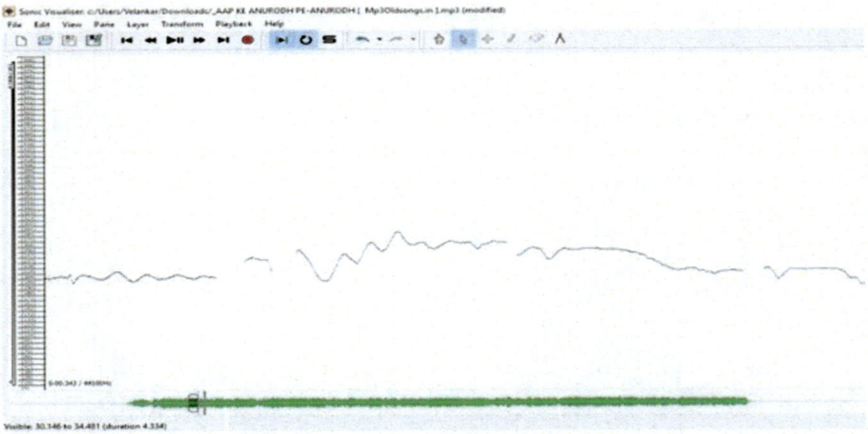

Fig. 3 Melodic pattern representation of pitch versus time

and applications of estimating F0 in monophonic and polyphonic recordings, as well as, the scope for research in this domain, have been discussed [42]. The free software and their add-ons listed in the paper are a useful resource for researchers.

We consider Sonic visualizer [45], a free and open-source feature extraction tool for the demonstration of a melodic pattern. In Fig. 3, the horizontal axis represents the timeline (selection of 4.334 s) and the vertical axis shows frequency in Hz (changes in the melodic pattern).

The pitch contour of a sound is a frequency function that tracks the perceived pitch of the sound over time [37]. As shown in Fig. 4, the pitch contour can be extracted using Praat software [38]. A just intonation filter bank is created using the tonic and its frequency value in the lowest octave. Pitch values are converted to represent the melody in notation form. Figure 4 shows the pitch contour of a small duration of one song from Geet Ramayana and pitch values extracted. Praat extracts pitch information in the most accurate manner for audio sample with speech or human voice in music [40, 51].

Instrumentation is the art of combining instruments in a musical composition (instrumentation). Extracting instrumental information from a mix of known instruments is a relatively easy problem compared to the actual problem of automatic extraction of instrumentation from a musical piece. It was attempted by employing deep neural networks [50].

Mel frequency cepstral coefficients (*MFCC*) are a set of 10–20 features that describes the overall shape of a spectral envelope. In music information retrieval (MIR), MFCC is often used to describe timbre. *Timbre* or tone is a perceived quality of sound that lets a listener feel the difference between the sounds produced by different sources. MFCC therefore has been employed in speaker recognition [4, 46]. MFCC seems a potential candidate for recognizing the instrumentation in a given piece of music [29].

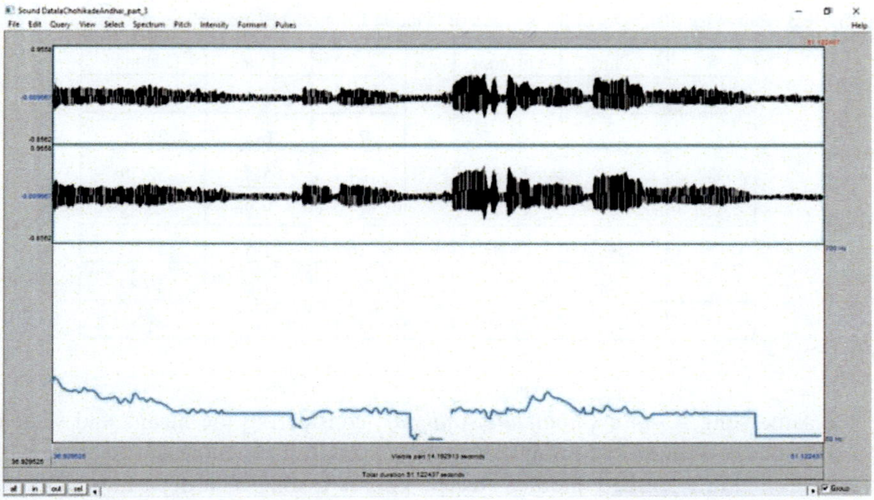

Fig. 4 Pitch contour using tool Praat

Contrast in general refers to a noticeable difference between two comparable things. Contrast in music refers to a change in rhythm, melody or harmony that is intended to increase the appeal of a composition and to provide a subconscious break to absorb the material from the main expository piece. It is a requirement for the aesthetic illusion of dramatic resolution of conflict. *Octave-based spectral contrast* embeds the information about both the relative spectral characteristics and the distribution of harmonic and non-harmonic components in each sub band [9]. Compared to MFCC, spectral contrast has a better discrimination potential for music types [26, 52]. In our observations the mean and standard deviations of a set of 12 MFCC and 6 contrast features are able to classify the 8 stanzas of a song from Geet Ramayana [8] with respect to instrumentation. The Euclidean distance between all pairs of the stanzas have been computed and produced in ascending order. Pairs of stanzas with similar features will be at the top of the list. The distances of the first pair and the last 7 pairs from the list are provided in Table 3.

The first three pairs, the first three rows in Table 3, form a cluster of stanzas 2, 3 and 4. The fourth record is a cluster of stanzas numbered 5 and 6. The 28th record shows that the instrumentation in the first stanza is much different than that in stanzas 6 and 7. These results are consistent with the observations [53] that the instrumentation in the stanzas 2, 3 and 4 consists of tanpura and violin while the rhythm has been added only for stanzas 5 and 7. There is very little instrumentation in the first stanza while the maximum number of sounds in the song is in the seventh stanza. The instrument in stanza 1 is a keyboard synthesizer while that in stanza 6 is a violin.

The *Chroma* features in music describe the tonal content of a musical audio signal in a condensed form [28]. Chroma could be employed to extract pitch-related information. The pitch has a direct relevance with the emotion to be conveyed [44]. To demonstrate this, we computed the Euclidean distances between the stanzas of

Table 3 Pair-wise distances of the stanzas of "Datala Chohikade Andhar"

#	Pair	Distance	#	Pair	Distance
1	2-3	7.81	22	1-3	26.21
2	2-4	10.37	23	6-8	30.06
3	3-4	11.64	24	5-8	32.79
4	5-7	12.42	25	7-8	34.88
5	3-8	16.07	26	1-5	37.01
6	2-8	17.74	27	1-6	40.82
7	4-7	18.69	28	1-7	41.21

the same song, *Datala Chohikade Andhaar*, considering the means and standard deviations of a set of 11 Chroma features. Stanzas 1, 2 and 6 formed a cluster while stanzas 3 and 4 formed another cluster. This is a song of a departing soul, king Dasharatha. In the first two stanzas he is recollecting the curse due to which he was to suffer the separation of his beloved son. In the sixth stanza the king in distress expresses that the curse has been realized and it is going to lead his family to an ever-worsening situation. So, the emotions in the three stanzas are much alike, while in stanza 3 the king says that there is no point in living when there is no Rama, his son, in his life. Next, in the 4 stanza he says that he cannot die in peace when he cannot see his Rama. So, the mood in both the stanzas is that the king is missing his son.

These short and limited examples are neither complete nor sufficient for a concrete mapping between the prosodic cues in a piece of music and the attributes of its digital signal [10, 24, 25]. However, they draw the readers' attention to a possibility that needs to be explored further with more and diverse datasets.

Below we present the details of our experiment of extracting musical information across the 56 songs of Geet Ramayana, by employing machine learning techniques.

Study of Prosody in Geet-Ramayana

Our database consists of the refrains of all 56 Geet Ramayana songs. The refrains are indexed S1 to S56, according to the sequence of the songs in the original book [13]. Clip size varies from 273 to 1036 KB and duration 9 to 34 s, leading to playing music at 30 KB per sec. However, this rate hardly has any correlation with the tempo of the music. The parameters of Chroma, MFCC and Contrast in the refrains have been computed using the Librosa library and the tempo has been computed by employing Logic Pro X. Details of the computations and observations are listed below.

1. All-pair Euclidean distances are computed for the Chroma, MFCC and Contrast Feature sets. Interestingly, the results of MFCC are identical with those of MFCC and Contrast, though Contrast alone led to a different clustering. The interesting

clusters generated with Chroma and MFCC features have been listed in the Table 4 along with the remarks of a musician.

Table 4 Cluster information

<table>
<tr><td>Cluster of similar songs</td><td>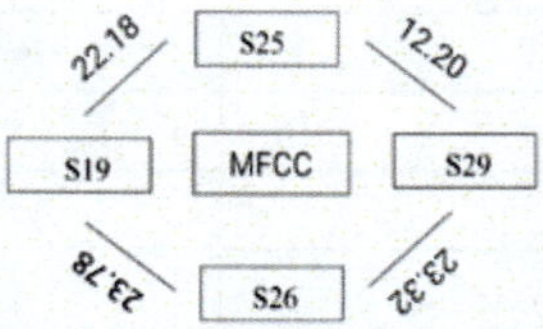
</td><td>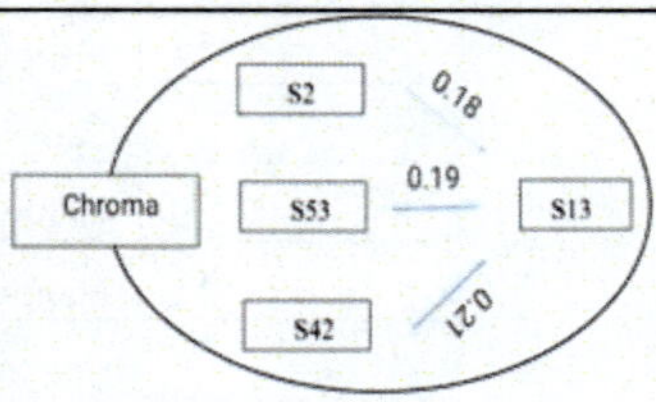
</td></tr>
<tr><td>Remarks</td><td>All four songs in the above cluster have drone, violin, harmonium and tabla. While S26 has hand cymbals, S19 has anklet bells in addition. The percussion instrument in 19 is Dholachi, that is different from Tabla. The distances between the nodes are consistent with this information</td><td>The mood indicated in all the four songs is, 'Better days are approaching'.
S13 is telling that the coronation of the hero has been declared. S2 describes the beautiful life in a city named Sharayu where the hero will take birth. S53 is seeking blessings and S42 is a team song while the construction of a bridge to reach the enemy is going on</td></tr>
<tr><td>Longest distance pairs</td><td><table>
<tr><th>Pair</th><th>Distance</th></tr>
<tr><td>S12-S50</td><td>230.79</td></tr>
<tr><td>S12-S40</td><td>225.20</td></tr>
<tr><td>S12-S27</td><td>214.70</td></tr>
<tr><td>S12-S20</td><td>211.73</td></tr>
<tr><td>S12-S26</td><td>210.22</td></tr>
<tr><td>S12-S22</td><td>210.14</td></tr>
<tr><td>S12-S4</td><td>209.13</td></tr>
<tr><td>S12-S29</td><td>208.46</td></tr>
</table></td><td><table>
<tr><th>Pair</th><th>Distance</th></tr>
<tr><td>S46-S22</td><td>0.93</td></tr>
<tr><td>S46-S43</td><td>0.84</td></tr>
<tr><td>S46-S55</td><td>0.79</td></tr>
<tr><td>S46-S21</td><td>0.79</td></tr>
<tr><td>S46-S19</td><td>0.78</td></tr>
<tr><td>S46-S41</td><td>0.77</td></tr>
</table></td></tr>
<tr><td>Remarks</td><td>S12 is a song of the wedding celebrations of the hero and heroine. Many instruments are used and the side rhythm is loud, while the songs that are far apart from it are indicative of anxiety, depression or a sigh of relief. Their instrumentation involves violin or synthesizer with Tabla. The big difference in the instrumentation has been rightly detected by MFCC</td><td>S46 describes the ultimate war that the hero is going to win; the mood in this song is very different than that in the others that are indicative of depression, anxiety and effort
Chroma parameters have successfully detected this difference</td></tr>
</table>

Table 5 Tempo of 20 refrains

Refrain ID	Mood of the song	Tempo
1	Sad towards end	60.82
6	Happy	62.56
12	Happy	76.9
3	Sad	81.1
10	Happy	97.65
11	Happy	102.67
2	Sad towards end	105.02
4	Sad	107.39
5	Happy	108.24
13	Happy	120.16
14	Anger	130.15
7	Happy	136.1
8	Fear	136.83
9	Anger	166.54

2. Tempos of the first 20 refrains in the Geet Ramayana set are provided in ascending order, as shown in Table 5.

Clearly, the tempo of a sad song is slower than that of a song with fear or anger. This is consistent with earlier findings [44]. However, a refrain of a song may be too small in order to conclude if the tempo of the song represented by it is rapid or firm. This prevents us from verifying the tempo-mood mapping whenever the value of mood is "Happy".

3. A number based upon the mood of a song was assigned to its refrain, assuming that −5 indicates extreme negative mood (generally, depression as a result of departure, loss, etc.), 0 indicates a philosophical mood (e.g., subscribing to "whatever happens was destined to happen") and 5 represents extreme positive mood (generally, the confidence of conquest). Figure 5 shows the trend in the mood-wise distribution of the refrains in the set. The peaks at −3 and 3 are respectively, due to the songs of departure and union.

As the data size is relatively small, we chose to employ regression-based ensemble learning to recognize the mood of a given song by considering its refrain as an input. We computed the mean and standard deviation of 12 Chroma features, 13 MFCC features and 7 Contrast features (feature index starts from 0) and also, mean, standard deviation and skew of Cent, Roll off and Zrate. Each record contains refrain id, song title, mood # (number) and these 74 features. Their interdependencies and the contribution of each of them to the identification of mood was estimated by computing the correlation.

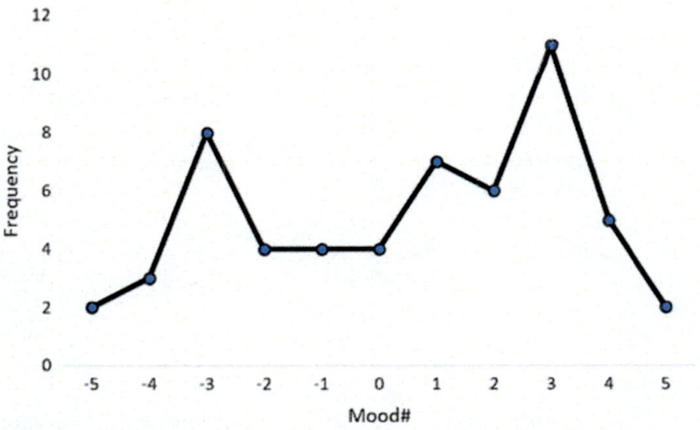

Fig. 5 Mood trends

In the following we provide a brief description of the three terms:

(i) *The cent* is a unit to measure musical intervals. It is 1/100 of the ratio between the two musical frequencies that are mapped to the adjacent keys on a 12-tone keyboard with equal temperament (Cent).
(ii) *Roll off* refers to the reduction of signal level when the signal goes outside the range of low bass and high treble, i.e., the two thresholds for low and high frequency sounds, respectively, that are audible to humans.
(iii) Z rate is the zero-crossing rate, i.e., the rate at which the audio signal changes direction from negative to positive or vice-versa. Generally, this feature is helpful in the classification of percussion sounds (zero crossing rate).

Preparation of training and testing datasets: The whole dataset was randomized. The first 15 records were separated to form a testing set. From the remaining 41 records we produced 5 bags by randomly selecting 15 records for duplication in each bag.

The interdependencies between the 74 features and contribution of each of them in the estimation of mood was estimated by computing their correlation matrix. Regression was carried out on each bag by selecting the 16 high-contributing features first and then by considering the remaining ones for reaching, by trial and error, an optimal performing configuration.

The parameters recommended by each of the models and their performance details are shown in Fig. 6.

The triplets in parentheses represent R-square, standard error and the highest value of p for the model. The pair in bold type, in brackets, is the percentage of records for which the model was able to identify the mood correctly in training and testing, respectively. The mood of a song being a fuzzy entity, we count it to be correct even if the machine computes the number one less or more compared to the number given by the expert.

The models that performed very well during training seem over-generated; they performed poorly during testing. The model that showed 66% accuracy has

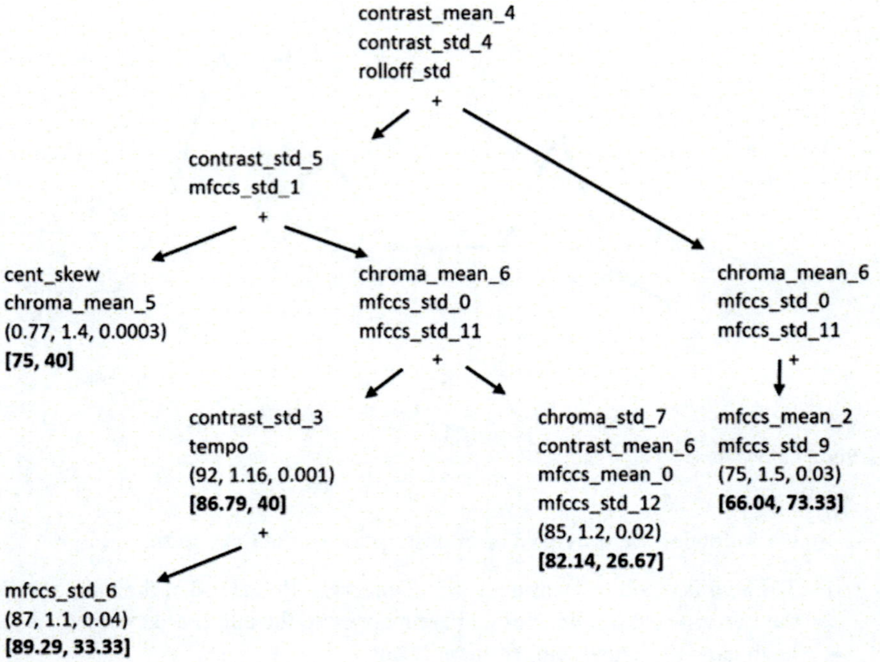

Fig. 6 Parameter modeling

performed comparatively well in the testing. Contrast 4 and Rolloff seem to be the common features across the models. Chroma 6 and MFCCS 0, 2, 9 and 11 have been proven to be important.

4. A confusion matrix of the misplacements observed in the training and testing of the best performing model has been computed as shown in Table 6.

Table 6 Confusion matrix of the misplaced records

Confusion matrix		Machine estimated mood#						
		−4	−2	−1	1	2	3	4
Human estimated mood#	−5					(0, 1)		
	−4		(2, 0)	(2, 0)				
	−3					(0, 1)		(0, 1)
	−2	(1, 0)			(0,1)			
	−1				(1, 0)	(1, 0)		
	0					(1, 0)	(1, 0)	
	1						(1, 0)	
	3				(1, 0)			
	4					(2, 0)		

The first and second elements of the ordered pair in Table 5 represent the frequency of misplacements when the model is tested with the training and testing data respectively. Clearly, there is no relation between the trend of misplacement at training and at testing.

Conclusion

The aim of this paper is to introduce some basic tools and techniques that are useful for estimating prosody in music. We have demonstrated them by the case of Hindustani classical music and by building a mood or sentiment classifier from a set of 56 Marathi songs. We hope that the experiments and results presented in this paper along with their limitations would attract researchers to take this work ahead. A prosody planning expert system would no longer be a dream.

Acknowledgements We sincerely thank Professor H. V. Sahasrabuddhe for sharing a topic for research and guiding us throughout. Thanks are due to Dr. Aranyakumar Munenni of University of Hyderabad for his indispensable role in this research. It is only after his critical evaluation of our results from the musical point of view that we ventured to publish this work. We acknowledge the sincere and timely support of Mr. Satyan Mumbarkar and Mr. Salim for their involvement in creating the refrain database of Geet Ramayana and bringing the concept of ensemble learning in this research, respectively.

References

1. E.M. Arroyo-Anlló, S. Dauphin, M.N. Fargeau, P. Ingrand, R. Gil, Music and emotion in Alzheimer's disease. Alzheimer's Res. Therapy **11**(1), 1–11 (2019)
2. F. Boutsen, Prosody: the music of language and speech. ASHA Leader **8**(4), 6–8 (2003)
3. A. Buxó-Lugo, Encoding and decoding of meaning through structured variability in intonational speech prosody (2019)
4. P.M. Chauhan, N.P. Desai, Mel frequency cepstral coefficients (MFCC) based speaker identification in noisy environment using wiener filter, in *2014 International Conference on Green Computing Communication and Electrical Engineering (ICGCCEE)*. IEEE (2014), pp. 1–5
5. M. Clayton, *The ethnography of embodied music interaction* (Routledge, 2017)
6. E. Coutinho, N. Dibben, Psychoacoustic cues to emotion in speech prosody and music. Cogn. Emot. **27**(4), 658–684 (2013)
7. A.S. Cowen, P. Laukka, H.A. Elfenbein, R. Liu, D. Keltner, The primacy of categories in the recognition of 12 emotions in speech prosody across two cultures. Nat. Hum. Behav. **3**(4), 369–382 (2019)
8. Datla chohikade song. https://www.aathavanitli-gani.com/Song/Datala_Chohikade_Andhar. Accessed in Feb 2019
9. F. De Leon, K. Martinez, Using timbre models for audio classification. Submission to Audio Classification (Train/Test) Tasks of MIREX (2013)
10. A. De Waele, A.S. Claeys, V. Cauberghe, G. Fannes, Spokespersons' nonverbal behavior in times of crisis: the relative importance of visual and vocal cues. J. Nonverbal Behav. **42**(4), 441–460 (2018)

11. M. Doffman, Groove: temporality, awareness and the feeling of entrainment in jazz performance, in *Experience and meaning in music performance*, ed. by M. Clayton, B. Dueck, L. Leante (Oxford University Press, Oxford, UK, 2013), pp. 62–85
12. H. Fujisaki, Information, prosody, and modeling-with emphasis on tonal features of speech, in *Speech Prosody 2004, International Conference* (2004)
13. G.D. Madgulkar, Geet Ramayana, Ministry of Information and Broadcasting, Government of India. ISBN 9788123019413 (2014)
14. Gadima. https://www.gadima.com/category/24/0/0/gadima-literature. Accessed in Feb 2019
15. Geetramayan. http://www.milindspandit.org/assets/GeetRamayana.pdf. Accessed in Feb 2019
16. D. Gibbon, Prosody: the rhythms and melodies of speech (2017). arXiv:1704.02565
17. A. Good, K.A. Gordon, B.C. Papsin, G. Nespoli, T. Hopyan, I. Peretz, F.A. Russo, Benefits of music training for perception of emotional speech prosody in deaf children with cochlear implants. Ear Hear. **38**(4), 455 (2017)
18. P. Grosche, M. Müller, Tempogram toolbox: Matlab implementations for tempo and pulse analysis of music recordings, in *Proceedings of the 12th International Conference on Music Information Retrieval (ISMIR)*, Miami, FL, USA (2011), pp. 24–28
19. W. Groß, U. Linden, T. Ostermann, Effects of music therapy in the treatment of children with delayed speech development-results of a pilot study. BMC Complement. Altern. Med. **10**(1), 1–10 (2010)
20. W. Gui, Y. Sun, Y. Tao, Y. Li, L. Meng, J. Zhang, A novel tempogram generating algorithm based on matching pursuit. Appl. Sci. **8**(4), 561 (2018)
21. M. Hausen, R. Torppa, V.R. Salmela, M. Vainio, T. Särkämö, Music and speech prosody: a common rhythm. Front. Psychol. **4**, 566 (2013)
22. G. Ilie, W.F. Thompson, A comparison of acoustic cues in music and speech for three dimensions of affect. Music Percept. **23**(4) (2006)
23. G. Ilie, W.F. Thompson, Experiential and cognitive changes following seven minutes exposure to music and speech. Music Percept. **28**(3) (2011)
24. A.P. Ismail, The usage of combined components of verbal L, vocal and visual (3-V components) of children in daily conversation: psycholinguistic observation. ELS J. Interdisc. Stud. Humanit. **2**(2), 290–301 (2019)
25. N. Jackob, T. Roessing, T. Petersen, Effects of verbal and nonverbal elements in communication. Verb Commun. 39–53 (2016)
26. D.N. Jiang, L. Lu, H.J. Zhang, J.H. Tao, L.H. Cai, Music type classification by spectral contrast feature, in *Proceedings. IEEE International Conference on Multimedia and Expo*, vol. 1. IEEE (2002), pp. 113–116
27. P.N. Juslin, P. Laukka, Communication of emotions in vocal expression and music performance: different channels, same code? Psychol. Bull. **129** (2003)
28. M. Kattel, A. Nepal, A.K. Shah, D. Shrestha, Chroma feature extraction, in *Conference: Chroma Feature Extraction Using Fourier Transform* (2019)
29. C.H. Lee, J.L. Shih, K.M. Yu, J.M. Su, Automatic music genre classification using modulation spectral contrast feature, in *2007 IEEE International Conference on Multimedia and Expo*. IEEE (2007), pp. 204–207
30. Y. Lévêque, L. Léard-Schneider, Perception of music and speech prosody after traumatic brain injury (2020)
31. J. Loewy, Integrating music, language and the voice in music therapy, in *Voices: A World Forum for Music Therapy*, vol. 4, no. 1 (2004)
32. A. Loutrari, M.P. Lorch, Preserved appreciation of aesthetic elements of speech and music prosody in an amusic individual: a holistic approach. Brain Cogn. **115**, 1–11 (2017)
33. E. Noth, A. Batliner, A. Kießling, R. Kompe, H. Niemann, Verbmobil: the use of prosody in the linguistic components of a speech understanding system. IEEE Trans. Speech Audio Process. **8**(5), 519–532 (2000)
34. J.L. Oliveira, M.E. Davies, F. Gouyon, L.P. Reis, Beat tracking for multiple applications: a multi-agent system architecture with state recovery. IEEE Trans. Audio Speech Lang. Process. **20**(10), 2696–2706 (2012)

35. C. Palmer, S. Hutchins, What is musical prosody? Psychol. Learn. Motiv. **46**, 245–278 (2006)
36. Patel, Peretz, Tramo, Labrecque, Processing prosodic and music patterns: a neuropsychological investigation. Brain Lang 61
37. Pitch contour. https://en.wikipedia.org/wiki/Pitch_contour. Accessed in Feb 2019
38. Praat. https://www.fon.hum.uva.nl/praat/manual/Intro_4_1__Viewing_a_pitch_contour.html. Accessed in Feb 2019
39. Prosody https://en.wikipedia.org/wiki/Prosody_(music). Accessed in Feb 2019
40. V.M. Ramesh, Exploring data analysis in music using tool Praat, in *2008 First International Conference on Emerging Trends in Engineering and Technology*. IEEE (2008), pp. 508–509
41. C.A. Ridgeway, J.M. Roberts, Urban popular music and interaction: a semantic relationship. Ethnomusicology **20**(2), 233–251 (1976)
42. J. Salamon, E. Gómez, D.P. Ellis, G. Richard, Melody extraction from polyphonic music signals: approaches, applications, and challenges. IEEE Signal Process. Mag. **31**(2), 118–134 (2014)
43. E.G. Schellenberg, S.E. Trehub, Frequency ratios and the perception of tone patterns. Psychon. Bull. Rev. **1**(2), 191–201 (1994)
44. K.R. Schere, Vocal affect expression: a review and a model for future research. Psychol. Bull. **99** (1996)
45. Sonic Visualiser. https://www.sonicvisualiser.org/. Accessed Feb 2019
46. Speaker Recognition. https://ccrma.stanford.edu/~orchi/Documents/speaker_recognition_report.pdf. Accessed Feb 2019
47. J. Tao, Y. Kang, A. Li, Prosody conversion from neutral speech to emotional speech. IEEE Trans. Audio Speech Lang. Process. **14**(4), 1145–1154 (2006)
48. W.F. Thompson, E.G. Schellenberg, G. Husain, Decoding speech prosody: do music lessons help? Emotion **4**(1), 46 (2004)
49. C.M. Tomaino, Effective music therapy techniques in the treatment of nonfluent aphasia. Ann. N. Y. Acad. Sci. **1252**(1), 312–317 (2012)
50. S. Uhlich, F. Giron, Y. Mitsufuji, Deep neural network-based instrument extraction from music, in *2015 IEEE International Conference on Acoustics, Speech and Signal Processing* (ICASSP). IEEE (2015), pp. 2135–2139
51. M. Velankar, A. Deshpande, P. Kulkarni, Music melodic pattern detection with pitch estimation algorithms (2018). https://doi.org/10.20944/preprints201811.0499.v1
52. K. West, S. Cox, Features and classifiers for the automatic classification of musical audio signals, in *ISMIR* (2004)
53. H. Sahasrabuddhe, Speech and demo during COMAD workshop at Mumbai University (2019)

Raga Recognition Using Neural Networks and N-grams of Melodies

Ashish Sharma and Ambuja Salgaonkar

Introduction and Background

Given a musical piece, the identification of its Raga is not a mundane task for a student of classical music but an intelligent problem. Let us call it RR, for raga recognition. Computer scientists have attempted to solve the problem by employing artificial intelligence. The process involves searching for raga-specific patterns in a given audio clip. Earlier it was done by signal processing [1, 2, 3], but now there is a trend is to analyze MIDI files [4]. (MIDI: short for musical instrument digital interface, is a music communication protocol for hardware and software interfaces [5]. MIDI files carry metadata, namely, notation, pitch, velocity, and tempo of the music, using which music could be recreated. Compared to other digital formats MIDI files are lighter and portable [6]. Obviously, they are more accessible and hence are widely preferred.) Along with statistical techniques of chromograms and histograms [7, 8], diverse machine learning algorithms, namely, K-nearest neighbors (KNN), support vector machine (SVM) [9], random forest [7], hidden Markov model (HMM) [10], pitch class distribution (PCD), and pitch class dyad distribution (PCDD) [11], latent Dirichlet allocation model (LDAM) [12], etc., have been employed for RR. Artificial neural networks (ANN) have remained the choice of several researchers [13, 4, 14]. A variety of datasets have been used. Their reported accuracies vary from 75 to 95%. A comparison of these works with ours has been provided in the Discussion Section of this paper.

More work seems to have been carried out in the domain of South Indian classical music [15, 13, 12, 16] while this research aims at Hindustani classical music (HCM), i.e., the North Indian variety. The novelty of this work is that just as with

A. Sharma (✉) · A. Salgaonkar
Department of Computer Science, University of Mumbai, Mumbai 400098, India

A. Salgaonkar
e-mail: ambujas@udcs.mu.ac.in

A. Salgaonkar and M. Velankar (eds.), *Computer Assisted Music and Dramatics*, Advances in Intelligent Systems and Computing 1444,
https://doi.org/10.1007/978-981-99-0887-5_7

humans, Aalap, Taan, and Bandish, the three types of melodic elaborations have been employed for learning the Raga-syntax by a machine. Information about note sequences has been stored in a library of unigrams (single notes), bi-grams (valid sequences of two notes), and tri-grams (valid sequences of three notes), upon which a simple feed forward ANN is trained. We propose to process notations, not audio signals. This makes our model simpler.

The model has been tested against HCM contained in three varieties: (i) unseen data of the population from which the training samples are drawn, (ii) songs based on the ragas, and (iii) live performances. The best of the performances of our model have demonstrated 100% recognition for one or two sets for five ragas out of the six in this study. An attempt has been made to optimize the ANN architecture. Descriptive statistics of these six ragas and their application to automatic music segmentation is a byproduct of this experiment. This also has revealed a heuristic for comprehending the nature of the ragas.

For readers who do not know HCM, we provide the basic terminology in the following section. In the closing part of this section, we reiterate the objective of this work. Next, in the section on materials and methods, we provide the details of data preparation and of building an ANN based classifier. The findings are further discussed in the results section. Contributions of this work and a few pointers for further research form the sections on discussion and conclusion.

Nomenclature: *Raga* is that which entertains and engages. Each Raga in Indian classical music is a melodic structure. It has unique well-defined *aroha* and *avroha,* ascending and descending scales. Each Raga is characterized by a distinguished set of melodic phrases known as *pakad* [17, 18]. A *bandish* is a composition in a raga. Elaboration and exploration of a composition within the framework of a raga is facilitated by *aalap*, and *taan*. An *aalap* is performed in a slow tempo while a *taan* has a faster tempo. Arguably, *aalap* and *taan* are sufficient to represent a rag. Light music and semi-classical music such as *thumri* and *gazal* are popular poetic genres of HCM that are more liberal in the exploration of the raga framework [19, 20, 21]. Natya Sangeet (theater music) and Indian film music are still more liberal forms of HCM [22, 23]. The perceived music is documented by employing a *notation* that consists of symbols to represent notes and their variants, duration of each, and durations of the absence of sound. Notations of the *aroha* that employ all 12 notes in Western music as well as HCM are shown in Table 1.

This paper aims at automatic RR of the following six ragas that are prescribed for an intermediate level course in HCM by a well-known music school: *Alhaiya Bilaval* (AB), *Bhairav* (BV), *Bhimpalasi* (BP), *Kedar* (KR), *Vrindavani Sarang* (VS), and *Yaman* (YN). Their ascending and descending notes are shown in Fig. 1.

RR from a signature group of notes is one of the expected competencies of these students. The experiment discussed in this paper is concerned with training a machine towards the same end.

Objective: To build a machine that correctly categorizes a given musical piece of a *line*, i.e., a sequence of around ten *phrases* (a small sequence of notes), in any of the six ragas.

Table 1 Western and HCM style notations

Notation Style	Note symbol
Western Music notation C# Scale	C# D D# E F F# G G# A A# B C
HCM notation (in Roman script)	S $\underline{R}$ R $\underline{G}$ G M $\overline{M}$ P $\underline{D}$ D $\underline{N}$ N
HCM notation (a variant in *Devnagri*)	सा रे॑ रे ग॑ ग म म॑ प ध॑ ध नी॑ नी

Note The '_' symbol in HCM notation (Roman) and '॑' in HCM notation (Devnagri), denote semitones, i.e., komal in the case of R, G, D, N and, sharp tone i.e. teevra in the case of M

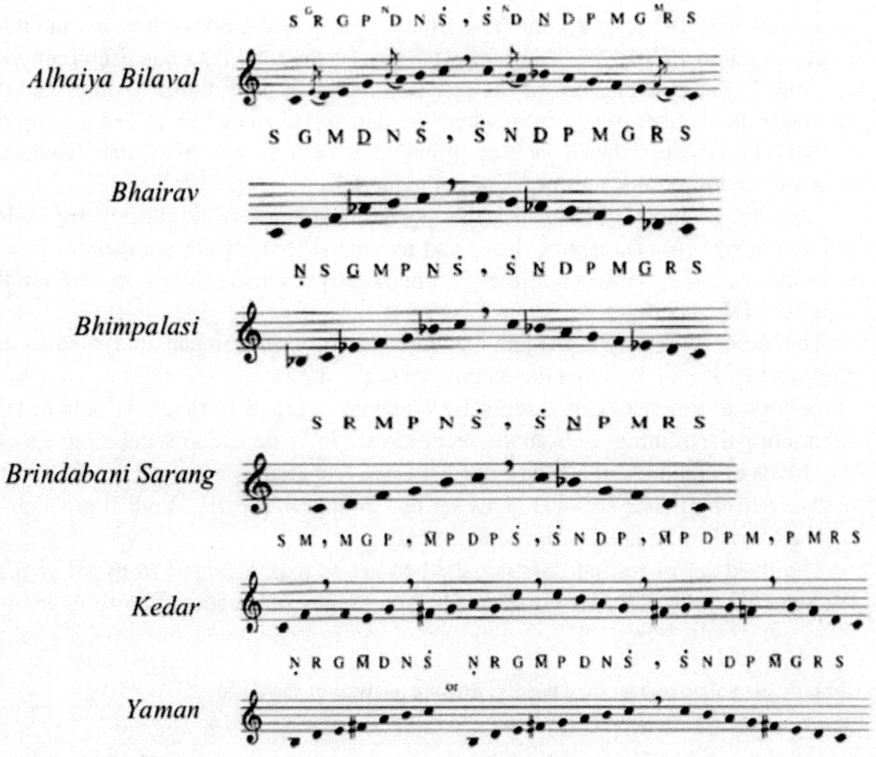

Fig. 1 Aaroh and avroh of the 6 ragas in our experiment (*Source*: 11)

Both the deterministic and probabilistic patterns of the ragas are captured by employing machine learning and natural language processing techniques on a set of *aalap*, *taan*, and *bandish* of the respective ragas. The input to our core system is generally obtained by automatic extraction of the notation information from a corresponding MIDI file or it is manually keyed in if the notations are available only in hardcopy form. The output is a name from one of the six ragas to which the input is closest.

The following section on material and methods presents details of data preparation and processing.

Material and Methods

Data

Training set is a database of the notations of *aalap*, *taan*, and *bandish* of the six ragas: AB, BV, BP, KR, VS and YN, from a standard text book of an intermediate level Course in HCM [24]. The raga-wise size of the total data has been compiled in Table 2. The first element of the pair in a record is the number of lines and the second is the number of sets from where the data has been gathered. The descriptive statistics of the actual data have been provided in Table 3. 80% of the data from each raga are randomly selected for training the model.

A record in Table 3 is a matrix that gives information of minimum size (Min), maximum size (Max), mean (Mean) and median (Med) for the number of lines in a composition (L), phrases per line (P), and notes per phrase (N) as observed in the samples of the six ragas.

The ratio of mean to median is close to unity. The mean and median are comparable. It indicates that the data is not skewed.

Testing data consists of three sets: The first set, call it H-set, is obtained by subtracting the training set from the set described in Table 2. Our second set is S-set. It consists of notations of 6 songs, one per raga, prepared by a trained musician after listening to the music [25–12]. This set has unique lines of 3 Hindi film songs, 1 from a Marathi film and 2 from Marathi plays.

The third set of testing data is called M-set as it is obtained from MIDI files. We selected the MIDI files corresponding to the six ragas according to their titles

Table 2 Size of aalap, taan, and bandish used for analysis of each raga

Raga	Feature			
	Aalap	Taan	Bandish	Total
AB	(26, 3)	(31, 4)	(17, 3)	(74, 10)
BV	(24, 3)	(34, 4)	(10, 2)	(68, 9)
BP	(17, 2)	(34, 4)	(27, 3)	(78, 9)
VS	(8, 1)	(35, 4)	(17, 3)	(60, 8)
KR	(19, 2)	(21, 2)	(12, 2)	(52, 6)
YN	(17, 2)	(27, 2)	(17, 3)	(61, 7)
Total	(111, 13)	(182, 20)	(100, 16)	(393, 49)

Table 3 Lines (L), phrases (P), and notes-wise (N) descriptive statistics for each raga

Raga		Descriptive statistics					Raga		Descriptive statistics				
		Min	Max	Mean	Med	Mean/Med			Min	Max	Mean	Med	Mean/Med
AB	L	3	13	7.4	7.5	0.99	BV	L	4	14	7.6	7	1.09
	P	1	28	12.5	13	0.96		P	3	24	9.2	9	1.02
	N	1.2	27	4.7	2.3	2.04		N	1.3	13.2	4.1	2.9	1.41
BP	L	4	14	8.7	8	1.09	VS	L	4	12	7.5	7.5	1.00
	P	1	34	13.4	14	0.96		P	2	32	12	12	1.00
	N	1.2	19.5	3.8	2.3	1.65		N	1.2	20.8	4.4	3	1.47
KR	L	4	15	8.7	8	1.09	YN	L	4	16	8.4	8	1.05
	P	4	24	11.2	11	1.02		P	1	20	10.2	11	0.93
	N	1	10.3	3.2	2.4	1.33		N	1.1	51	5.8	3	1.93

from a collection of 329 recordings of raga performances that were given to us by a musicologist. In the set there was no file named Bhimpalasi. But, there were two files named Bhim and Bhim2; we assumed that both were Bhimpalasi.

The lines in the S-set data were separated by taking cognizance of the song-lyrics. However, to compute the same in the M-set and further to separate the phrases in both the sets we resorted to descriptive statistics (Table 3). Though it depicts the individual raga characteristics about the phrases and line lengths, we in this pilot study go by a distribution suggested by their means (Table 4).

Hence, we obtain the details of S and M-Sets as given in Table 5.

Table 4 Overall lines, phrases and notes wise descriptive statistics

Data elements	Min	1st Qu	Avg	Med	3rd Qu	Max
Overall mean of L	4	6	8.7	8	9.2	15
Overall mean of P	4	9.5	11.2	11	13.2	24
Overall mean of N	1	5	3.2	2.4	5.3	10.3

Note Qu stands for Quartile

Table 5 Details of S-set and M-set

Raga	Particulars of S-Set			Particulars of M-Set	
	Song	#Lines	Source	Duration (min)	#Lines
AB	*Din Gele BhajanaviN*	*7*	[25]	2:56	24
BV	*Dil Ek Mandir Hai*	9	[20]	4:14	35
BP	*Indrayani Kaathi*	5	[21]	5:56, 7:39	50, 42
VS	*Kanta Majasi Tuchi*	7	[2]	4:25	35
KR	*Hum Ko Man Ki Sakti de*	10	[3]	2:39	36
YN	*Chandan Sa Badan*	*11*	[12]	4:20	37

Data modeling by employing an NLP technique: Music is understood as a human expression. Raga in Indian classical music conveys a mood [25] that is facilitated through the phrases for which specific notes and their sequences play a vital role. Not all the combinations of the set of notes of a raga are valid combinations. In other words, the raga-phrases follow a specific syntax. With this understanding, we propose a linguistic model for ragas. Notes in a raga form its alphabet, phrases are like words, and a composition is like a text.

N-gram analysis has been successfully employed to capture syntactic information in diverse scripts [27, 25]. A similar approach has been proposed here. The unigrams (single note patterns), bigrams (patterns in successive two notes) and trigrams (patterns in successive three notes) have been computed from the phrases in the data. Like the term-document matrix [28], a gram-raga matrix has been constructed. The (i, j)th element of this matrix denotes the frequency of the ith gram in the jth raga as observed in the training dataset. Figure 2 is a snapshot of the first few cells of the set of 7 g-raga training matrices created from *Aalap*, *Taan*, and *Bandish* and their combinations. The dimensions of all of them are given in Table 6.

Train A	Alaya E	Bhairav	Bhimp	Kedar	Vrinda	Yaman
ग	70	65	1	0	0	66
ग़	0	1	0	0	0	0
गं	11	4	0	0	0	11
ग ग	0	1	0	0	0	0
ग प	29	4	1	0	0	0
ग प ग	0	1	0	0	0	0
ग प म	3	2	0	0	0	0
ग म	5	48	0	0	0	1
ग़ म़	0	1	0	0	0	0
गं मं	1	2	0	0	0	0
ग म धं	0	7	0	0	0	0
ग़ म़ धं़	0	1	0	0	0	0
ग म नी	0	1	0	0	0	0
ग म प	0	5	0	0	0	0
ग म रें	0	16	0	0	0	0
गं मं रें	0	1	0	0	0	0
गं मं रैं	0	1	0	0	0	0
ग मरें	0	2	0	0	0	0
ग मरें र	0	1	0	0	0	0
ग रें	0	5	0	0	0	0
गं रैं	0	1	0	0	0	0

Train T	Alaya E	Bhairav	Bhimp	Kedar	Vrinda	Yaman
ग	208	182	5	7	0	214
गं	26	16	0	0	0	27
ग ग	7	8	0	0	0	12
गं गं	3	4	0	0	0	8
गं गं गं	0	1	0	0	0	1
ग ग रें	0	8	0	0	0	0
गं गं रैं	0	3	0	0	0	0
ग प	67	7	0	5	0	5
ग प प	7	1	0	0	0	0
ग प म	3	6	0	1	0	0
ग म	12	67	4	0	0	0
गं मं	5	1	0	0	0	0
ग म ग	9	7	0	0	0	0
ग म धं	0	3	0	0	0	0
ग म प	0	19	4	0	0	0
ग म म	0	1	0	0	0	0
ग म रें	0	4	0	0	0	0
गं मं रैं	0	1	0	0	0	0
ग मरें	0	4	0	0	0	0
ग मरें ग	0	2	0	0	0	0
ग मरें र	0	2	0	0	0	0

Train B	Alaya E	Bha
ग	19	
ग म	0	
ग रें	0	
धं	0	
धं धं	0	
धंन्	0	
धंन् धं	0	
नी	18	
नी सां	2	
नी सां रें	0	
न्	0	
न् धं	0	
न् धंन्	0	
न् धंन् धं	0	
न् सां	0	
न् सां सां	0	
प	28	
प ग	0	
प ग म	0	
प प	0	
प म	0	

Fig. 2 A snapshot showing a few cells of gram-raga matrices

Table 6 Seven gram-raga matrices and their dimensions

Matrix	Of the grams found in	Dimension
A	Aalap	398 × 7
T	Taan	705 × 7
B	Bandish	217 × 7
AB	Aalap and Bandish	483 × 7
TB	Taan and Bandish	782 × 7
AT	Aalap and Taan	802 × 7
ATB	Aalap, Taan and Bandish	869 × 7

Table 7 Dimensions of the gram-raga matrices of the testing data

#	Matrix	Dimension	#	Matrix	Dimension	#	Matrix	Dimension
1	A	398 × 7	4	AB	482 × 7	7	ATB	869 × 7
2	T	705 × 7	5	TB	782 × 7	8	S-set	482 × 7
3	B	217 × 7	6	AT	802 × 7	9	M-set	927 × 8

It is noted that though the *Bandish* and *Aalap* datasets are comparable in size (111 and 100 lines respectively), the number of unique patterns observed in the Aalap data are almost double the unique patterns observed in Bandish data. Though the Bandish dataset is the smallest amongst all, it has 67 out of 217 patterns that are new to Aalap and Taan datasets. This confirms that more difficult and aesthetic patterns are explored in Bandish.

Building a Machine Learning Classifier

For the high dimensionality of input data (gram-raga matrices) and the expected crisp classification (one label per record), we choose a supervised ANN-based classifier. A multinomial raga classifier has been constructed by employing a feed forward neural network (NN) with a single hidden layer. The model has been built by using the nnet package in R, a widely used open-source software for statistical computing [29]. Seven models are trained using the gram-raga matrices (Table 6).

Further, gram-raga matrices of the following dimensions have been constructed out of the testing data sets (Table 7).

Models are tested against the corresponding testing matrices, and the matrices of S-set and M-set are tested against each of the 7 models, thereby producing 3 sets of observations per model and 21 sets of observations in total.

Results

R0: The 7 models have varying impact on RR.

We test the following hypothesis to validate R0.

Hypothesis (H_0): *There is no statistically significant difference between the performance of the 7 raga classification models A, T, B, AT, AB, TB and ATB.*

Testing of Hypothesis: As there is no particular bias in choosing the samples, and all observations are independent, we employ the Kruskal–Wallis rank sum test [30] on the mean accuracies of the models as listed in Table 8; the corresponding Chi square test statistics has been given in Table 9.

Table 9 shows that at 95% level of significance, the p-value of the Kruskal–Wallis rank sum test is less than 0.05. Therefore, the null hypothesis is rejected. Therefore it is confirmed that at least one of the interactions dominates in the performances.

Table 8 Mean accuracies of various models on set H, S and M

Model	H-set (%)	S-set (%)	M-set (%)	Model	H-set (%)	S-set (%)	M-set (%)
ATB	93.14	91.84	72.90	*T*	77.28	59.18	70.75
TB	95.77	75.51	61.15	*A*	75.36	69.39	70.70
AT	81.85	73.47	73.18	*B*	72.22	74.45	70.25
AB	78.62	71.43	73.71				

Table 9 Kruskal–Wallis Rank Sum Test

Chi-squared Test statistic	Degrees of freedom	P-value
8.4378	2	0.01471

Next, we compute the percentage of correctly identified samples, mean accuracy, and the confusion matrix for each tested model. The details are provided in the next section.

R1: The performance of the RR models varies across the ragas and it also varies within samples of one raga.

The following Table 10 shows percentage of correctly labeled data of each raga and model. Models are listed along the rows and ragas along the columns.

Table 10 Percentage of correctly labeled data by each model and raga

	BP	BV	AB	YN	BS	KD
ATB	100.0, 100.0, 42.00, 92.86	100.0, 100.0, 82.86	100.0, 100.0, 62.40	84.62, 100.0, 89.19	83.33, 100.0, 71.43	90.91, 60.00, 69.44
AT	92.59, 60.00, 50.00, 76.19	100.0, 100.0, 77.14	82.35, 85.71, 100.0	58.82, 63.64, 54.05	82.35, 100.0, 74.29	75.00, 40.00, 80.56
TB	84.62, 100.0, 32.00, 66.67	100.0, 88.89, 57.14	90.00, 71.43, 79.17	100.0, 72.73, 100.0	100.0, 100.0, 51.43	100.0, 40.00, 41.67
AB	100.0, 60.00, 34.00, 80.95	85.71, 55.56, 74.29	88.89, 85.71, 83.33	71.43, 63.64, 100.0	40.00, 100.0, 62.86	85.71, 70.00, 80.56
B	100.0, 80.00, 85.71, 50.00	50.00, 77.77, 48.57	100.0, 57.14, 91.66	75.00, 81.81, 94.59	75.00, 100.0, 62.85	33.33, 50.00, 58.33
T	92.59, 60.00, 66.00, 80.95	100.0, 55.56, 57.14	76.47, 57.14, 66.67	58.82, 63.64, 91.89	94.12, 100.0, 74.29	41.67, 30.00, 58.33
A	96.30, 80.00, 36.00, 73.81	100.0, 88.89, 88.57	58.82, 42.86, 75.00	70.59, 90.91, 91.89	76.47, 85.71, 65.71	50.00, 30.00, 63.89

The first row in each cell gives the accuracy of our model on the H-set. The sample mean is identified with the population mean from where the training samples were drawn. 65% of the time it is above 80%, out of which 31% of the time it is 100%; the next figure in the cell indicates the accuracy on the S-set. 50% of the time it is above 80% out of which 29% of the time it is 100%. The overall performance of the models on S-set is comparable with their performance on the H-set.

The last figures in the cell are for the accuracy when the model is tested against the M-set. Recall that there were two MIDI files named Bhim and Bhim2, and we took both for Bhimpalasi. The model performed remarkably poorly with the first sample. However, the model's accuracy is 100% in 7% of the records of the other sample; in the same sample, it is above 80% in 38% of the cases and 78% of the time it is above 60%.

Arguably, the fall in the figures explains the flexibility of the raga framework that has generally helped musicians to create their stamp through their performances.

R2: The data about at least two among *Aalap*, *Taan*, and *Bandish* are required to build a satisfactorily performing RR model.

The following bar-chart (Fig. 3) depicts the mean accuracy of all 7 melodic interactions of *Aalap*, *Taan,* and *Bandish* while testing H-set, S-set, and M-set.

While TB is the best performing model on the H-set, it has average performance on the S-set and it performs the poorest on the MIDI file data. The model trained on ATB is the second-best performing model for the H-set. For the S-set it has the highest mean accuracy and for the M-set also it has a high mean accuracy. Therefore, ATB could be recommended as the first choice for building a classifier.

The AT and AB models have comparable performances while AT performs a little better for the H-set and AB performs a little better for the M-set.

The models trained by using a single feature are low performing. Therefore, it could be said that in order to build a satisfactory RR system, if not all of the three, we require data from Aalap with the data about at least one of the remaining two.

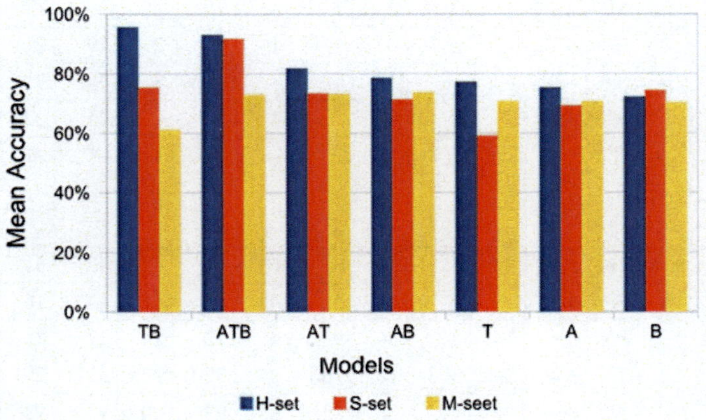

Fig. 3 Mean accuracy of the models on test samples

R3: Results of the three-feature model and that of the two-feature models are comparable though the three-feature model performs slightly better.

The confusion matrices of ATB and AB (relatively the least performing amongst the three), have been computed and their standard performance statistics has been shown in the last few rows of Table 11.

Abbreviations: True Positive (TP), True Negative (TN), False Positive (FP), False Negative (FN), Accuracy (in percentage) (AC), and F-Score (in percentage) (F).

Formulae:

$$AC = \frac{(TP + TN)}{(TP + TN + FP + FN)} \times 100$$

Table 11 Confusion matrices and performance metric of ATB and AB models

Train	→	AB		BV		BP		KR		VS		YN	
Test ↓↓	Set	ATB	AB	ATB	AB	ATB	AB	ATB	AB	ATB	AB	ATB	AB
AB	H	15	8	–	–	–	–	–	1	–	–	–	–
	S	7	6	–	–	–	–	–	1	–	–	–	–
	M	15	20	4	–	–	–	2	1	–	–	3	3
BV	H	–	–	14	6	–	–	–	–	–	1	–	–
	S	–	–	9	5	–	–	–	4	–	–	–	–
	M	1	–	29	26	2	2	1	–	1	4	1	3
BP	H	–	–	–	–	16	9	–	–	–	–	–	–
	S	–	–	–	–	5	3	–	1	–	1	–	–
	M	1	3	12	23	21	17	3	5	13	2	–	–
BP2	M	–	–	–	1	39	34	2	7	1	–	–	–
KR	H	–	1	–	–	–	–	10	6	–	–	–	–
	S	2	2	1	1	–	–	6	7	1	–	–	–
	M	2	1	3	2	3	–	25	29	–	–	3	4
VS	H	–	–	–	–	1	1	1	2	10	2	–	–
	S	–	–	–	–	–	–	–	–	7	7	–	–
	M	–	3	7	9	1	1	2	–	25	22	–	–
YN	H	1	1	–	–	–	–	1	1	–	–	11	5
	S	–	3	–	–	–	–	–	1	–	–	11	7
	M	1	–	–	–	2	–	–	–	1	–	33	37
Perfor-mance Metric	TN	37	34	52	37	81	63	41	42	42	31	55	49
	FP	9	6	6	14	32	43	15	11	12	16	6	6
	FN	8	14	27	36	9	4	12	24	17	8	7	10
	TN	334	298	303	265	266	242	320	275	317	297	320	287
	AC	96	94	91	86	89	87	93	90	93	93	97	95
	F	81	77	76	60	80	73	75	71	74	72	89	86

Table 12 Confirmation of the first sample of BP in M-set as an outlier

Sr No	View	# Experts	Designation
1	Bhimpalasi	2	1 beginner, 1 senior
2	Confused	3	1 beginner, 2 performing artists
3	Maybe Dhani	2	Senior professors in music
4	Jog + R	3	Professor and researcher

$$F = \frac{TP}{2TP + FP + FN} \times 200$$

The maximum misplacements, around 36 and 34% of the total, that are observed in the ATB and AB models are due to the first sample of Bhimpalasi (BP) in the M-set. The next smallest count of misplacement in both models is 14%, which is almost 39% less than that of the BP sample in M-set. Arguably, the BP sample is an outlier in this dataset. However, this needed confirmation.

R4: The first sample of BP in M-set cannot be confirmed as belonging to BP.

The experts were asked to listen to the clip and identify the raga without informing them about our doubt. Their views have been compiled in Table 12.

There seem to be 4 different expressions by 10 human experts whose field experiences vary from 2 to 30 years. All but one, seniors have opined that the clip cannot be easily labeled as Bhimpalasi. Clearly, the machine has successfully noted this difference.

R5: Removal of the outlier enhances the performance of the model.

We computed the Accuracy and F-score with and without the outlier. The results have been listed in Table 13.

The accuracy differs very little across the models. Also, removal of the outlier hardly improves the accuracy. However, F-score, a combined measure of precision (positive predictive rate) and recall (TP rate or sensitivity), shows around 8% rise, which could be considered as a significant improvement.

R6: AB is preferable to AT.

Ours is a 2-layer neural network with logistic activation function (Fig. 11); each perceptron of the hidden layer receives an input from each of the input nodes (number

Table 13 Model performance with and without the outlier record

Model	Accuracy			F-score		
	With outlier (%)	Excluding outlier (%)	% rise	With outlier (%)	Excluding outlier (%)	% rise
AB	91	91	0	73	79	8.2
ATB	93	94	1.1	79	85	7.6
% rise	2.2	3.3		8.2	7.6	

of unique unigrams, bigrams and trigrams identified for the model). The selected six ragas form the output layer of the six perceptrons (Fig. 4).

The number of perceptrons in the hidden layer is a parameter that drives optimization.

The optimal size of the hidden layer in each of the 7 combinations has been identified by trial and error. The models have been listed in decreasing order of the size of the hidden layer (Table 14).

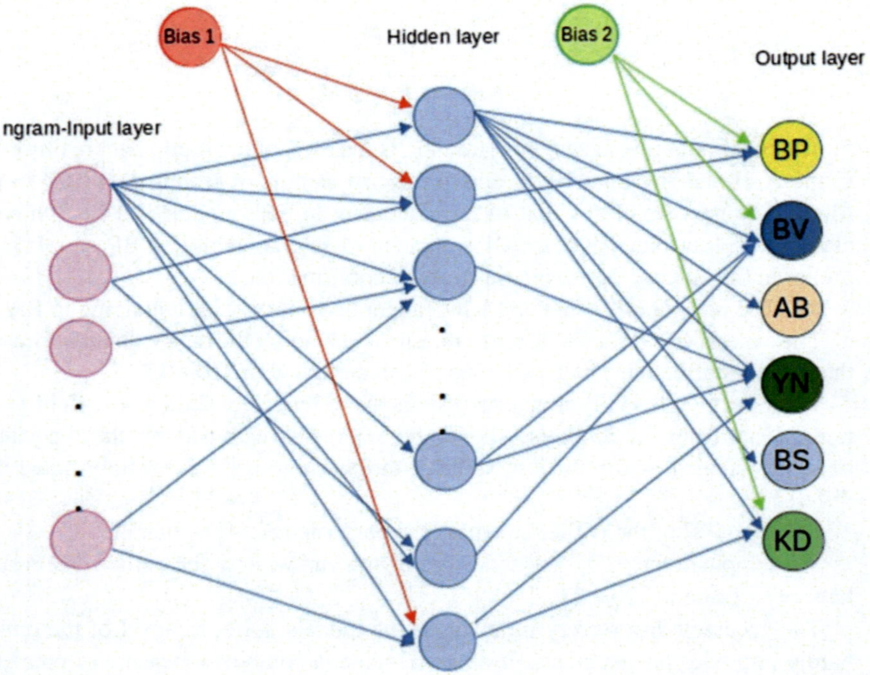

Fig. 4 Schematic of the ANN architecture

Table 14 Recommended size of the hidden layer

Model type	Size of the hidden layer (S)	Perceptrons in the input layer (I)	Min number of interactions S*I
A	10	398	3980
AB	9	482	4338
B	9	218	1962
TB	8	782	6256
AT	8	802	6416
T	6	705	4230
ATB	5	869	4345

Table 14 reveals the minimum number of interactions involved in each of the models for the optimal size of the hidden layer. This in turn is a measure of the complexity or cost of training the model. Accordingly, model AB is recommended over AT, as its cost is comparable with that of ATB while the cost of AT is about 50% higher.

Discussion

D1: Comparison of our model with that of the contemporary research.

Table 15 enumerates the results reported by recent researchers in increasing order of accuracy.

The present work is comparable with that of rows 4 and 5 in Table 15 in terms of genre, volume of the input data, objectives, and results. The novelty of our approach consists of employing n-grams as the input to the ANN. Unlike the others, our testing sample consists of the three varieties (H-set, S-set, and M-set). These two, along with the recognition of the outlier is an indication of the model's reliability. Also, the training set of our model has not been derived from audio performances but is based on the notations in a standard text book.

D2: What does the relatively low performance of our model on M-set suggest?

About 84% and 65% of the misplacements in the models ATB and AB respectively are due to M-set. Does it mean that the patterns in the live performances are deviating

Table 15 Results of recent work on raga recognition

Sr No	Technique, % accuracy, [reference]	Data
1	LSTM-RNN, 88–97%, [14]	Carnatic Music Dataset (CMD): 480 recordings of 40 ragas, audios of 124 h
2	Neural Network, 95%, [4]	MIDI files of 90 songs of 50 raga
3	ANN on aroha and avaraoha, 95%, [31]	training set: 50 audio files of raga testing set: 90 songs of 20 ragas
4	Random forest, structural analysis based on histogram of notes 94%, [7]	Audio files of 8 ragas: Bihag, Basant, Bhairavi, Darbari, Khamaj, Malhar, Sohoni, and Yaman
5	SVM, 92%; KNN, 87%, [9]	Recordings of live concerts, original CDs and downloaded audios of raga Bhairav, Bhairavi, Todi, and Yaman; tested on 60 wave files
6	Bayes multinomial, 92%, 1-nearest neighbour, 90%; Logistic regression, 85%; SVM with linear basis kernel, 81%; Naïve Bayes Bernoulli, 75%, [32]	CMD'

Table 16 Observations and possible explanations

#	Observation	Explanation
1	YN has longer phrases but have small numbers per line	The notes of YN offer scope for a larger variety of compositions. A longer phrase could form a line or, depending upon the artist's approach, a line could be made up of small phrases
2	Phrases of KR and BV are shorter in comparison with those of BP and VS. Also, lines of BP and VS are composed of more phrases	In contrast to YN or BP, BV and KR are tightly structured ragas. So, they offer less scope for variety. KR is less tightly structured compared to BV
3	Phrases and lines of AB are of moderate length	This may be due to its *vakra-chalan* (convoluted sequence). Articulation of phrases in vakra chalan is not as easy as it is in the straight chalan of a raga like Yaman

from the standard framework of the ragas? The acceptability of such performances needs to be explored. Then our training models could be upgraded accordingly.

Alternatively, the low recognition could be due to the error in the formulation of n-grams from these performances? Recall that in this pilot project we employed mean statistics in order to compute the phrases from the MIDI files, though the statistics had revealed assorted characteristics of the ragas. This choice was made with the assumption that the lengths of the phrases and lines in a performance are the choices of a performer and they are not dictated by the nature of a raga. However, later discussions with a few learned musicians revealed that though the artist has control over phrase length and line length, the aroha and avaroha of a raga have a significant impact on the improvisation of a raga. More details in this regard are provided in Table 16.

In the light of the arguments in Table 16, we imagine that the enhancement in the model is possible (i) by deploying raga-specific descriptive statistics in music segmentation and hence increasing the percentage of valid n-grams in the input matrix, and (ii) the n-grams in the lines that are misplaced could be evaluated by human experts and the input n-gram matrices could be updated accordingly.

The outcome of this research is summarized in the next section.

Conclusion

The objective of this work was to build an ANN-based raga classifier. We could successfully build it for a set of 6 ragas prescribed for an intermediate level examination. However, there are a few hundred popular ragas in HCM. We are aware that such small-scale experiments are hardly adequate for any generalization about a full-fledged raga-recognition system. More and in-depth studies are called for. However, the success of this experiment brings out the possibility of NLP and ANN-based raga classifiers.

We showed that (i) the data about *Aalap*, *Taan*, and *Bandish* are sufficient to build a raga classifier of HCM. The Kruskal–Wallis rank sum test reveals that each of these features and their combinations have different potential for building a raga classifier. Data about at least two of these features is required for building a satisfactorily performing raga classifier. (ii) This data is capable of giving a rough idea about the nature of the raga. These inputs may be useful for research in automatic segmentation of music. (iii) The n-gram based ANN raga classifier is able to identify outliers. This indicates its discriminating power to determine the differences or variations between ragas having identical sets of notes. The reported accuracy of this classifier on heterogeneous data sets is 94% and F-score 85%, which is quite satisfactory. (iv) Optimum size of the hidden layer varies for each of the models. The size drives the cost of the model and hence is useful in selecting a model for a given situation.

We claim that compared to the models that analyze the audio data for RR, our model is simpler because it processes the notations of a musical piece. The counter argument is that unlike western music, notation is neither native to HCM nor is it a driving force. A notation in HCM is just a hint for raga elaboration. It is often stated that a raga cannot be captured in the notations. Yes, we agree. Intonation, pauses and many other adornments play a vital role in making a raga presentation an aesthetic experience. Measurement of those parameters in a given musical piece will be necessary while evaluating the quality of a performance. This is altogether a different research problem. Designing an aesthetic scale for a performing art in general and for HCM in particular is a related problem.

In summary, from the datasets we tested for the selected 6 ragas, it is clear that our attempt of automatic RR has been helped sufficiently by the existence of raga specific basic phrases that are prescribed for an intermediate level student of HCM. More and exhaustive studies are required for a generalization of this assertion.

Acknowledgements We sincerely thank Shree Srijan Deshpande, Coordinator, Manipal-Samvaad Center for Indian Music, Dr. Aranyakumar of HCU, Dr. Manisha Kulkarni, Dr. Atindra Sarvadikar, Dr. Anuradha Garge, Dr. Ashwini Karwande and our students and friends in University of Mumbai for their continuous involvement in the evaluation of our models and generating inputs for its enhancement in music point of view. Our special thanks are due to Professor Hari Sahasrabuddhe for providing the MIDI database.

References

1. M.H. Bellur, V. Ishwar, X. Serra, A knowledge based signal processing approach to tonic identification in Indian classical music, in *Proceedings of the 2nd CompMusic Workshop* (2012), pp. 113–118 [Online]. Available: http://hdl.handle.net/10230/20423
2. R. Pendekar, S.P. Mahajan, R. Mujumdar, P. Ganoo, Harmonium RR. Int. J. Mach. Learn. Comput. **3**(4), 352–356 (2013). https://doi.org/10.7763/ijmlc.2013.v3.336
3. R. Sridhar, *Raga Identification of Carnatic music for Music Information Retrieval* (2009)

4. R. Sudha, A. Kathirvel, R.M.D. Sundaram, A System of Tool for Identifying Ragas Using MIDI, in *2009 Second International Conference on Computer and Electrical Engineering*, vol. 2 (2009), pp. 644–647. https://doi.org/10.1109/ICCEE.2009.49
5. D. Smith, C. Wood, The 'USI', or Universal Synthesizer Interface. Paper 1845, (1981). http://www.aes.org/e-lib/browse.cfm?elib=11909
6. https://en.wikipedia.org/wiki/MIDI
7. P. Dighe, H. Karnick, B. Raj, Swara histogram based structural analysis and identification of Indian classical ragas, in *Proceedings of the 14th International Society on Music Information and Retrieval Conference ISMIR 2013*, pp. 35–40, 2013.
8. P. Dighe, P. Agrawal, H. Karnick, S. Thota, B. Raj, Scale independent raga identification using chromagram patterns and swara based features, in *2013 IEEE International Conference on Multimedia and Expo Workshops (ICMEW)* (2013), pp. 1–4. https://doi.org/10.1109/ICMEW.2013.6618238
9. Y.H. Dandawate, P. Kumari, A. Bidkar, Indian instrumental music: Raga analysis and classification, in *Proceedings of the 2015 1st International Conference on Next Generation Computing Technologies (NGCT) 2015*, (2016), pp. 725–729. https://doi.org/10.1109/NGCT.2015.7375216
10. G. Pandey, C. Mishra, P. Ipe, Tansen: a system for automatic raga identification. Indian Int. Conf. Artif. Intell 1350–1363 (2003) [Online]. Available: http://citeseerx.ist.psu.edu/viewdoc/download?doi=10.1.1.60.8712&rep=rep1&type=pdf
11. P. Chordia, A. Rae, Raag recognition using pitch-class and pitch-class dyad distributions, in *Proceedings of the 8th International Conference on Music Information and Retrieval ISMIR 2007* (2007), pp. 431–436
12. R. Sridhar, M. Subramanian, B.M. Lavanya, B. Malinidevi, T.V. Geetha, Latent Dirichlet allocation model for Raga identification of Carnatic music. J. Comput. Sci. **7**(11), 1711–1716 (2011). https://doi.org/10.3844/jcssp.2011.1711.1716
13. J.C. Ross, A. Mishra, K.K. Ganguli, P. Bhattacharyya, P. Rao, Identifying raga similarity through embeddings learned from compositions' notation, in *Proceedings of the 18th International Society for Music Information Retrieval Conference (ISMIR)* (2017), pp. 515–522
14. S.T. Madhusudhan, G. Chowdhary, *Deepsrgm-Sequence Classification and Ranking in Indian Classical Music with Deep Learning*, pp. 533–540
15. H.G. Ranjani, S. Arthi, T.V Sreenivas, Carnatic music analysis: Shadja, swara identification and rAga verification in AlApana using stochastic models, in *2011 IEEE Workshop on Applications of Signal Processing to Audio and Acoustics (WASPAA)*, (2011), pp. 29–32. https://doi.org/10.1109/ASPAA.2011.6082295
16. S. Shetty, K.K. Achary, S. Hegde, Clustering of ragas based on jump sequence for automatic raga identification. Commun. Comput. Inf. Sci. **292**, 318–328 (2012). https://doi.org/10.1007/978-3-642-31686-9_38
17. T.M. Krishna, Carnatic music: Svara, Gamaka, Motif and Raga identity, in *Proceedings of the 2nd CompMusic Workshop* (2012), pp. 12–18 [Online]. Available: http://hdl.handle.net/10230/20494
18. S.X. Gulati, J. Serrà, K.K. Ganguli, S. Sentürk, Time-delayed melody surfaces for Rāga recognition, in *ISMIR 2016. ISMIR 2016. Proceedings of the 17th International Society for Music Information Retrieval Conference* (2016), pp. 751–757 [Online]. Available: http://hdl.handle.net/10230/33117
19. M. Nadkarni, A critical appraisal of thumri. J. Indian Musicol. Soc. **19**, 40 (1988) [Online]. Available: https://search.proquest.com/openview/ba6b8d93ed5703b925c79fcd350ed537/1?pq-origsite=gscholar&cbl=1816366
20. P. Manuel, A Historical Survey of the Urdu Gazal-Song in India. Asian Music **20**(1), 93–113 (1988) [Online]. Available: http://www.jstor.org/stable/833856
21. P. Manuel, The evolution of modern Thumrī. Ethnomusicology **30**(3), 470–490 (1986) [Online]. Available: http://www.jstor.org/stable/851590
22. M. Choudhury, R. Bhagwan, K. Bali, The use of melodic scales in Bollywood music: an empirical study, in *ISMIR* (2013)

23. V.M. Tilak, *Marathi Natyasangeet: Ek Swatantra Sangeet Prakar-Swarup Ani Samiksha.* S.N.D.T. Women's University (1990)
24. B.R. Deodhar, M. Raag-Bodh, Mumbai: Deodhar School of Indian Music (2012)
25. A. Mathur, S. Vijayakumar, B. Chakrabarti, N. Singh, Emotional responses to Hindustani raga music: the role of musical structure. Front. Psychol. **6** (2015). https://doi.org/10.3389/fpsyg.2015.00513
26. P.M. Sindhu, *Information encoding and network security applications of yantra images in manuscripts*. Ph.D. dissertation, Department of Computer Science, University of Mumbai (2020). http://hdl.handle.net/10603/317751
27. N. Yadav, H. Joglekar, P.N. Rao, M.N. Vahia, R. Adhikari, I. Mahadevan, Statistical analysis of the indus script using n-grams. PLOS One **5**(3), e9506– (2010)
28. W. Cavnar, J. Trenkle, N-gram-based text categorization, in *Proceedings of the Third Annual Symposium on Document Analysis on Information Retrieval* (2001)
29. B. Ripley, V. William, Feed-forward neural networks and multinomial log-linear models [package 'nnet']. CRAN, p. 4 (2020), [Online]. Available: https://cran.r-project.org/web/packages/nnet/nnet.pdf
30. M. Wolfe, D.A. Hollander, *Nonparametric Statistical Methods* (John Wiley & Sons, New York, 1973)
31. S. Shetty, K.K. Achary, Raga mining of Indian music by extracting Arohana-Avarohana pattern. Int. J. Recent Trends Eng. **1**(1) (2009). Accessed: May 03, 2011. [Online]. Available: https://www.researchgate.net/publication/228885219_Raga_Mining_of_Indian_Music_by_Extracting_Arohana-Avarohana_Pattern
32. S. Gulati, J. Serr, X. Serra, Phrase-based Raga recognition using vector space modeling music technology group, Universitat Pompeu Fabra, Barcelona, Spain Telefonica Research, Barcelona, Spain. Icassp **2016**, 66–70 (2016)

Developing a Musicality Scale for Haiku-Likes

Ambuja Salgaonkar, Anjali Nigwekar, and Atindra Sarvadikar

Preamble

Understanding *musicality*, the potential of a poem that turns it into a song lyric (song), is an essential skill for an aspiring musician. However, historically, unlike poetry and barring the folk literature, songs have neither been graded as *literature* nor have they been formalized [1]. Musicality is discussed as a part of music training rather informally and understood by the pupils intuitively by watching senior musicians. This may be also due to the fact that the musicality of a poem is partly dependent on the variety of music in which the lyric is going to be sung. However, across the varieties of music there seems an agreement about a few of the parameters of musicality. These include the existence of small events and things, short descriptions, repetition, simplicity, dialogue and authenticity [2] and the scope for inserting non-words or syllables as a part of improvisation [3]. The research reported in this paper pertains to the computational modelling of musicality for Hindustani classical music (HCM).

Though the lyrical requirements for presenting a raag (a melody framework in HCM) have been drastically changed over the long tradition of more than a thousand years, they have been understood by the practitioners of HCM in their own times. The trend of reduction in the length of the poem and the performance, over the last ten centuries, motivates us to work on a future variant that is comparable with Haiku,

A. Salgaonkar (✉)
Department of Computer Science, University of Mumbai, Mumbai 400098, India
e-mail: ambujas@udcs.mu.ac.in

A. Nigwekar
Department of Music and Dramatics, Shivaji University, Kolhapur 416004, India

A. Sarvadikar
Department of Music, University of Mumbai, Mumbai 400098, India

A. Salgaonkar and M. Velankar (eds.), *Computer Assisted Music and Dramatics*, Advances in Intelligent Systems and Computing 1444,
https://doi.org/10.1007/978-981-99-0887-5_8

a succinct Japanese poetry form of 17 syllables or so. We name the presentation Haiku-gaan, HG in short (the Sanskrit word gaan refers to singing activity).

It has been observed that the semantics of a lyric matters to a music composer, though it may sometimes lose its importance during the improvisation where consonance of notes takes priority in order to build an anticipatory mood within the audience. Haiku, a 3-line poetry originated in the seventeenth century Japan [4], has 5, 7 and 5 syllables respectively in its lines with at least one rhyme in it. It depicts a situation and a twist to it or it conveys a message [5]. This makes haiku a semantically rich and attractive poem that seems to satisfy many of the criteria for being a song.

Conversely, the *density of a haiku*, average number of letters used to describe an event in a literature, may demand a lot of imagination and understanding of culture in general on the part of its readers. Arguably, haiku set to music can be mutually complementary. It would convey the mood of the poem to listeners.

Clearly, not all the poems are songs or vice-versa; Haikus are no exception. The motivations and objectives of poem and song are not the same. Developing a device for computing their intersection has been a long-standing problem in literature [6].

Though the definitions of poetry forms from nursery rhymes to epics are available, defining poetry is difficult, or rather impossible. Separating poetry from prose has been a human skill [7]. Separating song lyrics from a collection of poems is still more difficult as they share the same form. A genius may be tempted to compose music for a beautiful poem even in the absence of so-called musical properties in it [8]. Consequently, musicality becomes a fuzzy entity.

Manual counting of phonological poetic devices, like figures of speech [9], alliteration, consonance, assonance and onomatopoeia, have been attempted [10]. Poetic speech melodies (PSM), newly discovered units that exist beyond syntactic phrases, lines or verses, form an interface between music and language and are measurable in terms of text-driven pitch and duration contours [11]. The paucity of available literature references in this domain indicates the scope for further research. Automatic computation of musicality of a given poem is one such a possibility that has been attempted in the present work. The work presented here is about developing a computational model to differentiate musical haikus from a collection of haikus in particular. It is assumed that the input to the system is a poem.

Nomenclature

Alliteration: The use of the same letter or sound at the beginning of words that are close together. Example: Bob brought a box of bricks.

Consonance: The sounds repeat, not letters. Example, Enough philosophy of fighting.

Assonance: The same vowel sound repeats within a group of words. Example, Not earth but the sun is a star.

Onomatopoeia: The words evoke the actual sound of the thing they refer to or describe. Example, Ding-dong of a doorbell.

Literature Survey

Computational creativity (CC) refers to creativity (i.e., the skill and imagination to make or do new things) by computer programs. CC aims at evolving a system by employing artificial intelligence so as to understand the concept of *creativity,* and thereafter, mimicking it. Synthesizing a meaningful text in a specific form like poetry is a creative art. Automatic creation of a poem has remained an ever-interesting topic of research in natural language processing. During the last decade, several machine learning models have been deployed successfully for the creation of haikus. A brief account of it is given below.

Hrescova and Machova [12] provides a comprehensive review of the poem models that play a role in automatic haiku creation. An extension of this work has been reported in [13]. Research described in [14] deploys a convolution neural network with long short-term memory for generating Haikus of type 5-7-5 and the estimation of their semantic and poetic value. Zhie [15] has employed the blackout poem approach, i.e., composing a poem by choosing relevant words from texts like newspaper articles. The computational aspects have been carried out by employing a natural language toolkit available in Python. This is the simplest approach in terms of the logic for haiku generation. It has passed the Turing test [16]. Many sequences of associated words matching a given theme are collected through several random walks of small distances in the English wordnet. The grammatically correct lines matching the Haiku templates are generated by filtering out the obviously non-desired entries by employing heuristics like avoid-word-repetition. 73% of the machine-generated haikus are found to be comparable with human articulated ones [17]. Reference [18] is an informally shared experience of writing a haiku generator with high acceptance rate using the *generative pre-trained transformer* [19]. It demonstrates the significance of choosing a large number of good quality haikus in the training database. While [20] discusses an image-based content creation and validation technique for generating haiku, [21] discusses a haiku generator that, by employing a database of traditional haikus, generates new ones and also voices them along with relevant sound and graphic effects by employing the *Internet of Things* [22]. Haikus are employed to demonstrate the worth of amalgamation of *genetic algorithms* [23] with a design framework for exploratory research in syntactic generators [24].

From the above it is evident that computational models are employed in generating and validating the Haikus. However, all these efforts are for Japanese and English poetry. No work has been found regarding poetry in any Indian language. The metric and modelling for poetry in Indian languages in general and Marathi haikus in particular may be fundamentally different because the concept of varna, the smallest unit of utterance that is used for counting word length, is analogous to a letter and not a syllable. We did not find any research on machine learning models for computing the musicality of haikus. Therefore, we claim that the choice of the theme and the dataset is a novelty of this research.

For setting up the required background of readers that are new to HCM, below we provide milestones in the evolution of HCM performance during the last one thousand years or so.

Evolution of HCM: Prabandha to Haiku-Gaan

In ancient days HCM artists would perform Prabandha, literally, a big essay [25]. In the twelfth century, it got reduced to eight-verse songs by Jayadeva. Next came Dhrupad with 4 verses. Later, as the focus shifted from devotion to romance, the Khayaal form was born. In its infancy during the eighteenth century, Khayaal was greatly impacted by the Dhrupad style; A few of the bandishes (literature) would be recited in both styles [26]. Over time Khayaal evolved as a precise and popular style of classical music. A Khayaal bandish would generally be of 2 stanzas (*sthayi* and *antara*). Consequently, musicality and thus the choice of words expressing sentiments and that were fit for music got preference in the selection or articulation of a bandish. It has been observed that the sounds cha, Na, bha, La, Sha, Tha, dha and kha (चणभळषठधख) rarely occur in the lyrics of a traditional bandish.

Over time the length standards of bandishes have changed. The smaller the bandish, the lesser it is driven by words and their semantics. This in turn facilitates musicians in retaining the required abstraction in the presentation of a Khayaal. Sometimes a bandish consists of a few lines of a poem. A few compositions of Kirana school of music consist of only one stanza, the sthayi [27]. Some modern artists have adapted the bandish form with only the first stanza, termed Asthai, as a part of their style.

HG in this proposal is an extrapolation of this trend, and musicality in our context is the potential of a poem, Marathi haiku in our case in place of a bandish, for setting music that facilitates improvisation or elaboration of a Raag.

Bandish, Musicality and Haiku in Indian Languages

Bandish, literature that is developed into a Raag-and-taal specific melodic composition and sung with rhythmic and melodic accompaniment [28] has remained an important and fascinating aspect of a standard performance in HCM. The notes of a Raag presented in slow tempo are called *Aalaap* and those presented in a faster tempo, *Taan*. A bandish is a locus for aesthetically developing these features of a Raag. The syllables of a Bandish facilitate adornments like Bol-alaap, Bol-taan and Lay-bol. The literal meaning of *bol* is a word or a meaningful sound. Bols have the potential to engage listeners even if they are not aficionados of classical music.

Because of the places where Khayaal evolved, Braja, a dialect of Hindi, remained the preferred language for the articulation of a bandish. Arguably, its choice instead of its allied languages, for example Hindi, is because of its musicality which is supposedly due to its simplicity. There are hardly any conjoint consonants. The dialect facilitates onomatopoeia, e.g., the word jhanana, the beginning word of the second line in the bandish below, describes the sounds of anklets. This naturally evokes in the listener memories of a young beloved.

Example 1 Braja-Bandish [29]

पायलिया (anklets)
झनकार (making a typical sound "jhan-jhan")
मोरी (mine).
झनन (jhanana, jhan-jhan in increased tempo)
बाजे झनकारी (leaves a mesmerizing musical impression of the sound jhanana ...).
पिया समझाऊँ समझत नाही (even if I explain, my beloved doesn't understand).
सास ननद मोरी देगी गारी (my mother-in-law, sister-in-law will abuse me).

Human relations, devotion, nature, seasonal beauty, philosophy, etc., have been the subjects of some all-time popular Bandishes. These have been the subjects of modern haikus as well.

Arguably, the appreciation of Khayaal will reach greater heights if the melody is built around a meaningful poem. The repertoire of traditional bandishes is sufficiently rich that musicians generally have a choice of presenting different bandishes at different performances and establish stylistic variations through their presentations. Contributions to the database of songs may enrich the taste and delight of the listeners. Could haikus handle this challenge?

Materials and Methods

Experiment 1: The objective of this experiment is to check the ease or difficulty of haiku making.

On July 5, 2018 we conducted a half-day Haiku-making workshop for MSc students and faculty at University of Mumbai. 40 students and 10 Professors participated. As an outcome of this session, we received 2 Tamil, 2 Hindi, 1 English, and 15 Marathi Haikus contributed by 6 teachers and 10 students. These people had moderate level exposure to the literature in their mother tongue. It was the first creative writing experience for our students. Among them, 4 teachers and 3 students have contributed 800 haikus on 50 odd topics during the last 3 and 1/2 years. Is an award-winning collection of 250 haikus. *Chitra-Haiku*, a collection of 200 pictures complemented by haikus, is ready for publication. The importance of these creations is that most of them are articulated on the given topics. Our experience is that the skill of articulating a Haiku-ish 3-liner (call Haiku-like) on a given topic could be picked up by young students with a little initial effort. Further, the skill keeps sharpening over time (Fig. 1).

This characteristic is more promising compared to the standard sigmoid-shaped learning curve that is generally expected for learning a new skill like the creation of

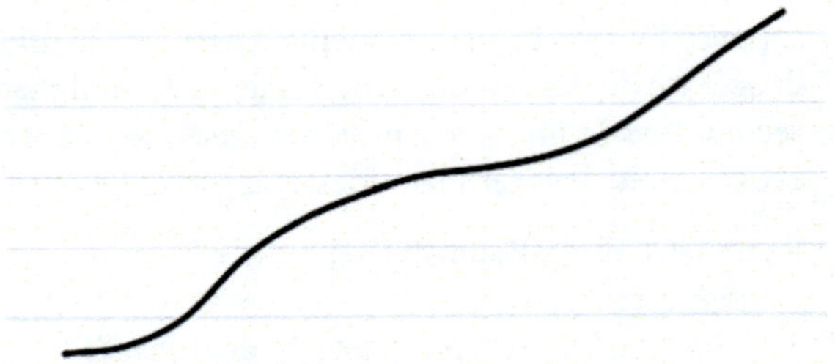

Fig. 1 Learning curve for composing a Haiku-ish poetry

poetry, for which the creation of a Khayaal bandish (a poem of 2–3 stanzas) is no exception.

So, the conclusion is that the creation of the required types of haiku-likes is not difficult. However, we reiterate that not all Haiku-likes would be candidates for Haiku-gaan. Therefore, research like the present one has its own worth.

The credit of bringing haiku to Indian languages goes to Ravindranath Tagore, though his haikus do not comply with the restriction on the letter counts of the lines. The same applies to the haikus of Shirish Pai, who introduced this form to Marathi [30].

Experiment 2: The objective of this experiment is to explore the possibility of setting haikus to music.

For this experiment, we selected 4 haikus in Indian languages from online resources. The fifth haiku was articulated as a BH by an accomplished musician. We requested four open-minded professors in music to rate and review them. The haikus and the subjective opinions of the experts have been listed below in Table 1.

The conclusion is that some of the Haiku-likes are musical and some more could be made musical by tweaking them. Concord in the poem matters.

Maybe because of the more frequent usage of aspirated consonants, the bandishes in Marathi and other Indian languages didn't become popular. Haiku being a tiny composition, requires a very limited number of letters. Our experience is that given

Table 1 Sample haikus in Indian languages and their ratings

Sr. No.	Language, haiku and the experts' ratings	Meaning in English	Remarks
1	(Hindi) छिड़ा जो युद्ध रोएगी मनुजता हसेंगे गिद्ध – Jagadeesh Vyom (5, 4, 4, 3)	The severity of war is explained War has commenced humanity will cry while vultures will be happy …	Uncomfortable with words like war and owl in a bandish. Dissonance reduces musicality. Simple, easy-to-pronounce words are required. More musical words need to be identified for मनुजता and हसेंगे

(continued)

Table 1 (continued)

Sr. No.	Language, haiku and the experts' ratings	Meaning in English	Remarks
2	(Marathi) डोळा बदामी डोहाचा निळाशार तरता चंद्र – Ambuja Salgaonkar (4, 5, 3, 2)	It is a simile A small blue pond with the moon's reflection reminds us of almond-shaped eyes. What do they stand for? Such interpretations are left to the reader	Conjoint consonants and anusvara at some places are a hindrance. Replace चंद्र with चंदर उन्हात doesn't sound musical in its place. The language is not an issue (Marathi Bandishes were first introduced in 1927 [44]). Easy to pronounce words would be preferred though difficult words are acceptable. In general, semantic complexity is inappropriate
3	(Marathi) डोई जळण तान्हेली गवळण तप्त उन्हात – ShahajiDhende (3, 4, 2, 3)	Difficult life of the underprivileged ones has been depicted A village woman is walking on the roads to sell buttermilk, a cold-drink; She herself is thirsty under the hot sun…	
4	(Braja) बिछुडे कैसे सिंधुसे जलधार प्यार अपार – Rameshwar Kamboj (2, 2, 2, 2)	This is an example of anthropomorphism How could water go away from an ocean? It is the love between the two that brings them together	Replace प्यार with प्रेम, जगे रे with जगा दे, euphony carries value, Braja haikus would be better candidates for BH, perhaps because even junior musicians are used to the musicality of Braja words, including ones with aspirated consonants We may be sceptical about the acceptance of the bandishes in the languages other than Braja
5	(Braja) हो निरंजन सबन के जीवन ज्योती जगे रे – Atindra Sarvadikar (1, 1, 1, 2)	These are auspicious thoughts Let everybody's life be pure Let a light be lit	

a situation or mood, a trained haiku poet suggests or articulates a haiku that consists of letters of the user's choice and also onomatopoeic words.

In the light of experts' comments, the next experiment is to set a few of the Haiku-likes to music according to time-of-day conventions called raag samay chakra in HCM.

Experiment 3: The objective is to compose Haiku-likes in HCM by following raag samaya chakra.

A century ago Marathi students of HCM were taught sant poetry and its transcreation as a bandish [44]. (*Sant poetry* consists of devotional or religious verses, or advice for the masses). The issue of dissonance was ruled out by construction. We transcreated *Doha*, sant poetry written in Braja and its close dialects of Hindi [32], to Marathi Haiku-like. In particular, a set of 52 *Doha-haikus* (call it DH) was derived from the Dohas of Sant Kabeer. Two senior musicians were invited to set them to music. Three samples from DH are listed in Table 2.

Table 2 Doha haikus

Sr. No.	Doha	Haiku
1	साई इतना दिजिए जा में कुटुम समाय मै भी भूखा न रहू, साधू न भूखा जाय	मजपुरते- दे आप्तांही पुरते इतुके दे चि
2	गुरू कुंभार, सीस कुंभ है, घडी घडी काडै खोट अंदर हाथ संवर दे, बाहर मारे चोट	गुरू कुंभार काढी शिष्यात खोट आधार देत
3	ऐसा कोई ना मिले, हमको दे उपदेस. भौ सागर में डूबता, कर गहि काढै केस.	गुरु भेटेना भवपाश तुटेना काय करावे

The observation is that 26 out of 52, i.e., 50%, DH were rated as BH. Also, a suggestion was made to combine more than one haikus to form a bandish. For example, the last two haikus of Table 2 are in praise of a teacher, the second describes the characteristic of identifying errors in his disciples and and removing them, while the third one is the longing of a disciple for a teacher. They could be elaborated as Sthayi and Antara.

Raag music was composed for 16 haiku-likes such that given a time of day, at least one composition would be available for recitation. During this process our understanding about musicality got improved. The outcomes of this experiment are compiled in Table 3.

The results of the experiment indicate that a sufficient number of haiku-likes from DH are musical can become haiku-bandishes (HB).

A Confirmatory Test and Conclusions

A collection of 25 randomly selected haikus from DH was offered to a group of trained young musicians in a formal gathering and they were invited to set one, two or three poems of their choice to music and present their compositions to an audience. They were asked not to select the same haikus, if possible. In a short span of one and half hours, 6 persons presented 17 compositions after selecting 12 different haikus.

The case of two persons choosing the same haiku occurred twice and three persons choosing the same haiku occurred once. The music composed by them was not exactly the same but it was clearly driven by the sentiment of the text. For example, the fifth DH in Table 3 was composed in Shankara by one of the artists and the twelfth in Puria Dhanashree. The composers' perceived intents were, "bringing clarity to the mind" and "compassion for the self" respectively; most observers would concur.

Interestingly, agreement in the selection of the junior and senior musicians was observed in 80% of the cases. Out of the three poems that were not in the set selected

Table 3 HB-Raag mapping (Raag information [33, 34])

Sr. No.	Doha-Haiku	Meaning	DH-sentiment (DHS)	Raag (anticipated sentiment (RS))	Time (RT)
1	दुःखी स्मरशी सुखात सदा स्मरे दुःख न उरे	Remembering the Almighty in good times saves us from misery	A balanced thought	Bhairav (maturity)	Sunrise
3	करिसि कधी प्रलय उद्या जर? आजच कर	You may not see tomorrow. Work today on your dreams	Awakening	Vibhaas (maturity, peace)	Sunrise
2	मजपुरते- आप्तांपुरतेही दे इतुके दैचि	Please give me and mine, just enough …	Abundance is assumed	Ahir Bhairav (peace)	Morning
4	कुटिल कोण? माझ्याइतका कोण? शोधितो व्यर्थ	No outside evil can be as cunning as the one within me My search is in vain	Introspection	Jaunpuri (deep devotion)	Late morning
5	या, सूप होऊ सत्व ते टिकवू फेकूनी फोल...	Let's be a winnower, collect grain by throwing away the chaff	Easily conveyed serious message	Gaud Sarang (devotion, romance)	Early afternoon
6	युद्ध पेटले राम की रहमान मर्म अज्ञान	Ignorance causes religious wars	Jealousy, Dominance	Vrindavani Sarang (disturbance)	Afternoon
7	धीर धरी रे नको ओतू पाणी रे फळ मोसमी...	Have fortitude. Don't panic. The fruit will appear in its season	Advice	Shuddha Saranga (maturity)	Late afternoon
8	गुण हो हिरा पारखी जर खरा ना तरी कौडी	A diamond doesn't get appraised unless it reaches a jeweller	Praise	Bhimpalasi (maturity, devotion)	Pre-evening
9	यमग्रास तू ताटात वा मुखात भ्रमी सुखात	An ignoramus enjoys life by excluding the thought of inevitable death	Indifference or anguish for death?	Shri (deep devotion, anxiety laden)	Sunset
10	येतती जात उमलत-सुकत उदयास्त-त्व...	Exit follows entry; rise follows fall by the law of nature	Parting	PuriaDhanaashri (maturity, compassion)	Sunset

(continued)

Table 3 (continued)

Sr. No.	Doha-Haiku	Meaning	DH-sentiment (DHS)	Raag (anticipated sentiment (RS))	Time (RT)
11	जन संपती जसे विरूनी जाती तारे प्रभाती	People vanish like the stars in the morning…	The truth	PuriaDhanaashri (maturity, compassion)	Sunset
12	रे बुडबुड्या नसू तू न मी उद्या क्षणभंगुरा...	Our existence is only for a while, like a bubble we shall vanish	Reminder of one's duty	Jayajayavanti (duty, achievement at its price)	Late evening
13	तळ धुंडाळू मोती मिळे हाताला वारा काठाला	Dive deep for the pearl; Sitting on the bank will get you a breeze	Deep message in lighter words, creativity	Hemant (depth, calm)	Night
14	वाणी रत्नचि तोलू-मापून बोल वक्त्याते मोल	Speech is a gem Speak precisely. Speakers are respected	Unnoticed wealth	Gaud Malhaar (romance/trick by disjuncts beloved)	Late night
15	न साबु-पाणी निंदक हो शेजारी स्वच्छता न्यारी	Critics in the neighbourhood are like soap and water; They bring clarity	Cleaning	Shankara (clarity)	Post mid-night
16	हंस वेचिती लाटेवरले मोती बक निर्बुद्ध	Swans pick up pearls as the waves roll in; egrets cannot	Mystical	Bhairavi (existence of the Supreme)	Any time

by the seniors, one was considered by altering two words while retaining the Haiku form. Revisiting the DH, the seniors and juniors together concluded that a few more haikus could be considered as BH. They also found that in many cases musicality of the haikus could be enhanced easily by modifying one or two words.

Empirically we confirm that 50% of the DH are so musical that even junior musicians could successfully compose classical music to present them. Now the final test is to check the appeal of HG performances.

Experiment 4: Objective of this experiment is to verify the attractiveness of HG. This was carried out in two stages. First, we tested a prototype by setting 5 DH to Raag music, and rendered them in a Master's degree music class at Shivaji University, Kolhapur. All in the audience appreciated the HG. A special appreciation was received for the lyrics being tiny and in their mother tongue. In the light of the fact that the traditional bandishes have been composed by the maestros, these students were excited by the possibility of reciting a bandish articulated by themselves.

Table 4 HB-Raag mapping (Raag information [33, 34])

Rating	DH-sentiment	Raag-sentiment	Time
−2	Dominance, jealousy	Disturbance	Afternoon
−1	Anguish, indifference, parting and unnoticed merit	Anxiety, compassion, distancing	Late night, sunset
0	Cleaning, duty, introspection, serious, truth	Clarity, devotion, duty	Excluding the times given above and below
1	Abundance, advice, awakening, balance, creativity, mase, mystical, praise	Ease, depth, supreme power, maturity, peace, romance, soothing	Sunrise, morning, night and independent-of-time

In the next phase, we trained a machine learning model by using the inputs of the musicians that participated in this series of experiments. The findings of a few analyses have been provided in the beginning of the next Section.

Computational modelling of musicality:

1. A model to discover the latent relationship between DHS, RS and RT.

Starting with the mapping between RS and RT that has been traditionally accepted [35], we associated each of the three factors DHS, RS and RT with a 4-point scale, by using a heuristic: a negative sentiment would be rated with −1 or −2 depending on its intensity, neutral sentiment would take value 0 and positive sentiment, 1. Times of day are rated according to the expected energy levels within us. The time at which people feel high energy levels is given a high rating. The lowest rating is for the time when people feel exhausted. The exact mapping is given in Table 4.

Correlation between the three has been computed by using the records in Table 5. These results are consistent with the experts' inputs that the choice of the Raag is driven by the semantics of the lyrics, DH in this case. The strong correlation between DH-Sentiment and Raag-time confirms that at any given time of day we have at least one HB suitable for HG as expected in the construction.

2. The various conjectures proposed by the musicians throughout this journey have been validated.

C1: The musicality of a DH is dependent of the choice of the letters.

Table 5 Correlations between DH and Raag Sentiments and Time

Correlations	DH-sentiment	Raag-sentiment	Raag-time
DH sentiment	1.00	0.81	0.87
Raag sentiment	0.81	1.00	0.81
Raag time	0.87	0.81	1.00

Table 6 Results of the t-tests over the letters in the Bandish and Haiku-likes in DH sets

P-values of T-tests	HB	Non-HB	U-HL
Bandish	0.37	0.39	0.17
HB	1.00	0.61	0.72
Non-HB	0.61	1.00	0.74

R1: Given four sets, namely, Marathi bandishes in [44], HB, Non-HB and Haiku-likes in DH of whose the musicality has not been tested manually (U-HL), the p-values of the two-tailed, paired t-test over the letter frequencies have been given in Table 6.

The probabilities ranging from 0.17 to 0.74 indicates that there is a high chance that the observed dissimilarity between the four sets is due to coincidence. It means there is no significant difference in the distribution of letters in the four sets.

C2: Occurrences of the less frequent letters in the songs reduce the musicality of a Haiku.

R2: The letter frequencies in the first 5% songs of a Marathi song database (call AG) [36] revealed that the letters अ, क, ग, च, ज, ड,, ण, त, द, न, प, म,, य, र, ल, ळ, व, श, स,corresponding to a, ka, ga, cha, ja, Da, Na, ta, da, na, pa, ma, ya, ra, la, La, va, sha, and sa of the Roman alphabet (call the M-set) form 80% of the total of the letter occurrences, while the letters ख, छ, ट, ठ, थ, फ, ष, घ, झ, ढ,ध, ब,i.e., the letters corresponding to kha, Cha, Ta, Tha, tha, pha, Sha, gha, jha, Dha, dha, ba and Bha (call the L-set) contribute to 14% of the letter occurrences. The frequency of the 19 letters in the M-set is 18 times higher than the frequency of 13 letters in the L-set.

However, the p-value of the two-tailed, paired t-test of the L-set, the M-set and the set of vowel frequencies in HB and non-HB in DH are 0.82, 0.74 and 0.99, respectively. These indicate that the musicality of Haiku-likes has not been driven by the choice of letters.

C3: Appearance of conjoint consonants affects musicality.

R3: The p-values of the two-tailed, paired t-test over the frequencies of the conjoint consonants in Bandish, HB, Non-HB and U-HL sets have been given in Table 7.

The probabilities ranging from 0.17 to 0.96 indicate that there is a high chance that the observed dissimilarity between the four sets is due to coincidence. It means that there is no significant difference in the frequencies of the conjoint consonants in the musical and non-musical texts.

C4: Rhyme and rhyming patterns have an impact on musicality.

Table 7 Comparison of the frequencies of the conjoint consonants in the Bandish and Haiku-likes

P-values of T-tests	HB	Non-HB	U-HL
Bandish	0.17	0.57	0.51
HB	1.00	0.43	0.29
Non-HB	0.43	1.00	0.96

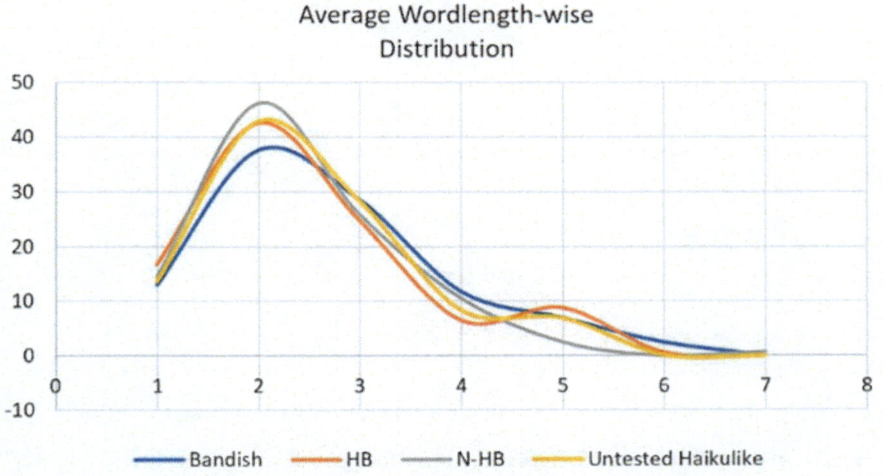

Fig. 2 Length-wise word frequency distribution in Bandish and Haiku-likes

R4: The p-value of the t-test of the samples of rhyming patterns in both HB with Non-HB and U-HL are 0.8 and 0.5, respectively, and that of Non-HB with U-HL is 0.34. This indicates that the observed difference between the sets is not statistically significant. Therefore, rhyme pattern too does not provably impact the musicality of the HB.

C5: Small words are preferred.

R5: The average word-length-wise distribution of words in different databases has been displayed in Fig. 2.

There is a slight difference between the distributions of N-HB and the others at word-length 5. It could be a clue for a feature leading to musicality.

C6: Alliteration carries positive weight while composing music for a poem.

R6: Let alliteration factor (AF) be the ratio of the number of unique letters to the total number of letters in a given text. Devanagari being a phonetic script, the AF covers the existence of consonance and assonance as well. The smaller the AF, the more is the repetition of letters in the text. The AF of Bandish, HB, Non-HB and untested Haiku-likes ranges from 0.29 to 0.49, 0.44 to 0.8, 0.5 to 0.77 and 0.44 to 0.88, respectively.

The AF-wise percentage of compositions in each category shown in Fig. 3 indicates that the AF of most of the bandishes is significantly less than that of the Haiku-likes. This result is consistent with the result reported in [11].

The distribution of letters in HB and untested Haiku-likes is not much different. This indicates the possibility of 50% of them having the AF required for them to be HB.

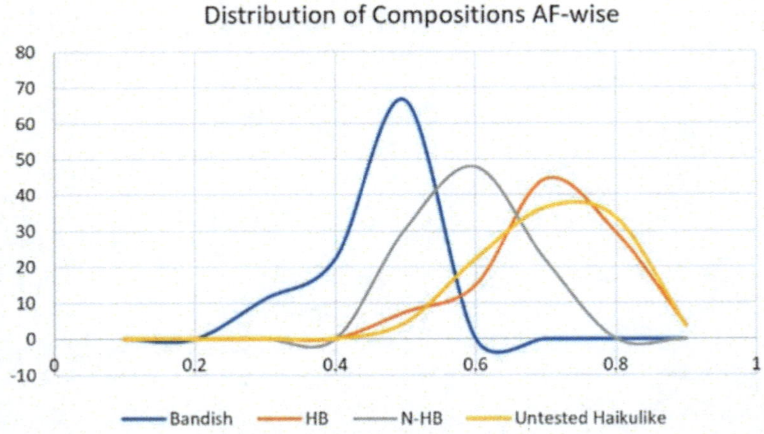

Fig. 3 AF-wise distribution of compositions in the four groups

C7: The long notes Aa (आ) and I (इ/ई), words beginning with Aa, and lines ending on a long note provide the scope for elaboration, and therefore, add to the musicality.

R7: The p-values of the t-test of the frequencies of I, Aa, words beginning with Aa and lines ending on a long note are compared below for all four sets (Table 8). The entries corresponding to the sets that are significantly different are italicized.

Appearance of the long notes and Aa seem to have a significant impact in comparison with the note I.

In short, starting with the experts' attempts at expressing their feelings, we analysed several features. Out of them, it appears that longer words, repetition of letters, and positions of the long notes affect musicality. Therefore, we select them as the features for building our machine learning model for a musicality scale.

3. Deciding the musicality of a haiku in this context is modelled as a binary classification problem.

3.1 Regression, supervised clustering and ensemble learning approach.

Table 8 Comparing the frequencies of I, Aa in Bandishs and Haiku-likes

P-Values of t-test	I (इ/ई)	Aa (आ)	Words beginning with Aa	Lines ending on a long note
Bandish and HB	0.18	*0.03*	0.20	*0.001*
Bandish and N-HB	0.24	0.50	*0.05*	*4.4E-05*
Bandish and U-HL	*0.04*	0.86	*0.05*	*1.6E-06*
HB and N-HB	0.83	*0.08*	*0.01*	0.65
HB and U-HL	0.88	*0.01*	*0.01*	0.53
N-HB and U-HL	0.67	0.56	0.27	0.95

Because it is a binary classifier, *logistic regression*-based modelling is a natural choice. The performance of the model in the test set was not found to be consistent with that of the training set. Therefore, we employed *principal component analysis* during pre-processing for feature reduction, and further, independently attempted *supervised k-means clustering* based training with the same datasets. The strengths of the regression model and the clustering models were found to be different, and that led us to the thought of an ensemble learning approach.

First we present the logic involved in these techniques and then compute the confusion matrices to provide the metrics of their performance.

Logistic regression (LR): Regression is a technique of estimating a relation between a set of predictors that are independent variables and a response that is a dependent variable. A regression line is a line from which the data points are located at optimal distances with respect to a predefined mathematical criterion, say least square sum. If for a given data set, instead of a line we get an S-shape curve that adheres to this property, then that curve is called a logistic regression curve representing or modelling the given data points. Linear regression is characterized by a linear equation, $Y = m_1X_1 + m_2X_2 + \ldots + m_kX_k + c$. In logistic regression Y is transformed to ln(P/(1-P)), where P is the probability of occurrence of a particular label in the given data. Linear regression is for a continuous variable Y, while logistic regression is for a discrete variable Y [37, 38].

Principal component analysis (PCA): The linear combinations of the independent variables are formed such that the contributions of the correlated variables are compacted by using orthogonal transformations so that the new variables are uncorrelated. The process involves the computation of the (i) inter-feature covariances of the standardized values of the independent variables, (ii) eigenvalues of the covariance matrix, (iii) eigenvectors corresponding to a few of the top eigenvalues and (iv) multiplication of the transpose of the eigenvectors with the transpose of standardized values of the independent variables. Computation of PCA and eigenvectors are explained in [39] and, [40] respectively.

Supervised K-means clustering (KMC): This is a technique for partitioning the data such that the records of similar characteristics are grouped together. In its unsupervised version, starting with a random partition, a stable grouping is achieved by letting the elements join an appropriate group. During each iteration the centroids of the current groups and the distance of each element from these centroids are computed and each element is put into a group whose centroid is the nearest to it. These inter-group movements of the elements cause shifts in the centroids [41]. In the supervised variant, the values of the response variable are provided as background knowledge. This leads to a faster and unique convergence.

Ensemble learning approach (ELA): It is a recommendation for seeking a better performance by combining multiple models when one or a few models show strengths in a few aspects but not in totality [45].

Data preparation: For some of the dohas we had articulated more than one haikus. However, only the musically best one among them was subjectively selected for D-H. At this stage we included the other variants as well. Implicitly, we equipped the machine with comparative information about musicality. The number of haikus in

Table 9 Performances of the diverse machine learning models for deciding musicality of Haikus

Observation template↓ \ Models →		LR	PCA + LR	KMC	ELA (LR+KMC)
Machine: +ve / -ve; Human +ve: **TP** / **FN**; Human -ve: **FP** / **TN**	Training	8, 7 / 4, 14	9, 6 / 8, 10	10, 5 / 9, 9	7, 8 / 2, 16
	Testing	6, 5 / 7, 8	6, 5 / 6, 9	7, 4 / 10, 5	6, 5 / 5, 10
Performance					
Precision (+ve predict): TP / (TP+FP)		(.67, .46)	(.53, .50)	(.53, .41)	(**.78**, **.55**)
Recall (True +ve rate): TP / (TP+FN)		(.53, .55)	(.60, .55)	(**.67**, **.64**)	(.47, .55)
Accuracy: (TP+TN) / (TP+TN+FP+FN)		(.67, .54)	(.58, .58)	(.58, .46)	(**.70**, **.62**)
F-Score: 2TP / (2TP+FP+FN)		(**.60**, .50)	(.56, .52)	(.59, .50)	(.58, **.55**)

our dataset thus became 59. We employed random stratified sampling over this set such that 60% of the data was for training and the remaining was for testing. Further, if an HB was included in the training set or the test set, then its variants were also included in those sets. As a result 56% of the samples were included in the training set and 44% in the test set.

We considered a 5-tuple: (% of lines ending on a long note, % of the words starting with Aa, % of the words with length 5, % of Aa, % of I), for modelling a haiku. The training and test datasets in this case consisted of 33 and 26 records, respectively.

Our observations of training and testing various models have been articulated in *confusion matrices*. The corresponding *performance metrics* have been provided in Table 9.

R1: The PCA in the pre-processing has shown a marginal rise in the performance of test data.

R2: In the training phase LR is found significantly poor in the recall parameter while KMC is found better in it.

R3: The ensemble of LR and KMC was expected to demonstrate an overall improvement. However, in testing it outperformed in all metrics but the recall (bold in the Table). In this ensemble the candidates for which there was no consensus between the labelling computed by the two models are labelled as N-HB.

The accuracy and F-scores of ELA with the testing data are 62% and 55% respectively. They indicate that there is a big scope for improvement in the model. The metamodel used in the above machine learning experiments is *bag of symbols* which does not hold the information about the context or location of the tokens. The question arises as to whether they define the syntax of a musical poem. Our next experiment was to find out an answer to this. We included the sequential information about the long and short notes in the data. The details of it are listed below.

3.2 A rule-based system to evaluate the musicality predicate over a given haiku.

A rhythmic cycle in HCM has two parts, namely, from Sam to Khaali and from Khaali to Sam. The beginning of a simple bandish can be aligned with respect to the beginning of either of them. However, the end of the bandish has to be aligned with respect to the end of the second part. This fact is interpreted as the starting and ending patterns of a bandish following special musical features with the ending patterns having more choices.

In Marathi the letters with the modifiers aa, ee and oo and also a letter preceding a conjoint consonant are to be pronounced in double the time for the modifiers a, i and u. We transformed haikus into binary sequences by replacing long letters with 2s and short ones with 1.

By scanning the first and the third lines of the HB and N-HB from the training set we created databases of the starting and ending patterns (SP and EP) and their respective frequencies. Out of 64 possibilities in each case we got 20 unique pattensin SP and 17 in EP. A musicality heuristic, HM, was articulated as follows:

Let *S_Pattern* and *E_Pattern* be the beginning and ending patterns of a candidate haiku (CH) whose musicality is to be tested. A *Pattern* is a binary sequence of long and short notes. It is either an S_pattern or an E_Pattern of a CH.

If(OR(!found_in(Union(SP, EP)), AND (Type(Pattern) = S_Pattern, Found_in(SP(N-HB)), !Found_in(SP(HB))), AND (Type(Pattern) = E_Pattern, Found_in(EP(N-HB)), !Found_in(EP(HB))))(Pattern) = 1, Label(CH) = N-HB, Label(CH) = HB).

A database update heuristic (UH) has been articulated:

If Label(CH) = HB then increase the frequencies of its starting and ending patterns respectively in the databases SP(HB) and EP(HB) by 1; Else, increase the frequencies in SP(NHB) and EP(NHB) by 1.

The assumption in this rule-based system is that there is no error in the labelling of the training set. Therefore, the confusion matrix and performance have been computed only for the test set [Table 10].

Like clustering here too the highest value in the performance metrics is recall. The recall of the rule-base system presented here is 100%, which means all the haikus that were labelled by humans as musical have also been labelled as musical by the machine. It labelled 25% of the records as musical, which the musicians

Table 10 Results of the Rule-base system for evaluating the Musicality predicate for a haiku

Confusion matrix

Testing Rulebase		Machine	
		+ve	-ve
Human	+ve	11	0
	-ve	4	11

Metric	%↑
Precision = 0.74	34.6
Recall = 1.0	56.3
Accuracy = 0.85	37.1
F-Score = 0.85	54.5

Upgrading Databases	Pre		Post		% rise	
	SP	EP	SP	EP	SP	EP
Common	7	10	8	10	14	0
In_HB	4	3	7	7	75	133
In NHB	9	4	9	4	0	0

Table 11 Results of the models in the perspective of the selection of a non-musical pattern

Observation template ↓ Models →	Rule-base	LR	PCA + LR	KMC	ELA
Machine: -ve / +ve Human -ve: **TP** / **FN** Human +ve: **FP** / **TN**	Training → Testing↓	14 4 7 8	10 8 6 9	9 9 5 10	16 2 8 7
	11 4 0 11	8 7 5 6	9 6 5 6	5 10 4 7	10 5 5 6
Precision	(-, **1.0**)	(.67, .63)	(.63, .64)	(.64, .56)	(**.67**, .67)
Recall	(-, **.73**)	(.78, .53)	(.56, .60)	(.50, .33)	(**.89**, .67)
Accuracy	(-, **.85**)	(.67, .54)	(.58, .58)	(.58, .46)	(**.70**, .62)
F-Score	(-, **.85**)	(.72, .57)	(.59, .62)	(.56, .42)	(**.76**, .67)

can verify. This could be interpreted as a liberal approach for selection of a poem for composing music. The significant growth in the database of musical patterns after testing demonstrates the potential of the model for discovering newer musical patterns through testing. Convergence of this phenomenon could be a pointer for further research.

A conservative approach towards the solution of this problem could be to select a system that performs better in identifying the non-musical haikus. The confusion matrices from this point of view are presented in Table 11.

Clearly, the rule-base has outperformed the rest, while ELA was ranked second.

We attempted to generate the results by including data about Marathi bandishes. To an extent we succeeded in getting a comparable performance in the models other than the rule-base. We could not continue with them in the rule-base model as the basic assumption of the beginning and ending phrases was based upon the haiku meter that doesn't match with the text of a traditional bandish. Creating the binary sequences of the traditional bandishes and annotating their beginning and ending phrases is a challenge that could be attempted by significant amounts of manual efforts in the future and the validity of this rule-base could be tested for the more generalized domain.

Conclusion

The predicate of musicality has been attempted by employing the machine learning techniques, namely Regression, PCA, Clustering, Ensembled and Rulebase. The model has been trained and tested over a set of tiny poems that is consistent with the trend of shorter lyrics in Hindustani classical music performances. The musicians' intuitions are validated succinctly. The best demonstrated accuracy and f-scores of the model with the proposed characterization of poetry for checking its musical values are 85%

While this research is getting concluded the number of DH has reached 108. More students are showing interest in composing music for them. The automatic creation of Hiaku-likes has become a topic for a few of our students. It is clear that the novelty of this research has motivated people of multiple disciplines to explore the domain. The pointers provided for interdisciplinary projects have also been a contribution of this work.

This is the beginning of research in this problem. There is scope for performance enhancement as well as generalization of the musicality scale proposed in this work.

Acknowledgements The authors take this opportunity to thank Professor H. V. Sahasrabuddhe for the seed concept of transcreation of Kabir-Dohas in Marathi haiku and presenting them musically in order to overcome the hindrance in their interpretations due to succinctness of the poetry. Thanks are due to Professor Vivek Patkar, Professor Jayant Kirtane, Dr. Aranyakumar and Professor Srijan Deshpande for their guidance and involvement throughout the progress of this research. We acknowledge the inputs of the students of the Department of Computer Science, University of Mumbai for their involvement in learning the articulation of Haiku-likes. Special thanks are due to Ms. Prachi Teli for writing a Python script that generates the letter-level descriptive statistics of the DH. The involvement of the students of Department of Music and Dramatics at Shivaji University, Kolhapur, deserves a special credit for their sustained interest for the enhancements while we were coming up with a set of musical compositions of DH.

References

1. M. Zapruder, The Difference Between Poetry and Song Lyrics. https://bostonreview.net/forum_response/difference-between-poetry-and-song-lyrics/
2. A. Stolpe, How to Write Song Lyrics. https://online.berklee.edu/takenote/how-to-write-song-lyrics/
3. https://www.britannica.com/art/song
4. https://www.britannica.com/art/haiku
5. S. Varma, Japani Haiku Aur Adhunik Hindi Kavita, Diamond Publications
6. H. Plooy, Narratology and the study of lyric poetry, Literator (2010)
7. https://www.britannica.com/art/poetry
8. Avaliya Khaliya, Samvaad, Editorial@Maharashtra Times, March 18 (2018). https://maharashtratimes.com/editorial/samwad/hridaynath-mangeshkar/articleshow/63338890.cms
9. https://www.litcharts.com/literary-devices-and-terms/figure-of-speech
10. H. Tawfiq, A Study of the Phonological Poetic Devices of Selected Poems of Robert Browning and Alfred Tennyson, English Language and Literature Studies, Journal of Canadian Center of Science and Education (2020)
11. W. Menninghaus, V. Wagner, C. Knoop, M. Scharinger, Poetic speech melody: a crucial link between music and language. PloS One (2018)
12. M. Hrescova, K. Machova, Michiko: Poem models used in automated haiku poetry creation, in *Proceedings of International Conference on Current Trends in Theory and Practices in Informatics* (Springer, 2017)
13. M. Hrescova, K. Machova, Haiku poetry generation using interactive GA and Poem template. Acta electromecanica et informatica (2017)

14. J. Neiman, Generating haiku with deep learning (Part I) (2018). https://towardsdatascience.com/generating-haiku-with-deep-learning-dbf5d18b4246
15. S. Zhie, NLP With Python: build a haiku machine in 50 lines of code—a dive into natural language processing (2002). https://betterprogramming.pub/nlp-with-python-build-a-haiku-machine-in-50-lines-of-code-6c7b6de959e3
16. https://en.wikipedia.org/wiki/Turing_test
17. Y. Netzer, D. Gabay, Y. Goldberg et al., Gaiku: generating haiku with word associations norms, in *Proceedings of the NAACL HLT Workshop on Computational Approaches to Linguistic Creativity* (Association for Computational Linguistics, Colorado, 2009)
18. (2019) https://www.brianweet.com/2019/06/16/write-ai-gpt-2-haiku.html
19. https://en.wikipedia.org/wiki/GPT-2
20. P. Davis, R. Kapistalam, I. Lindsay et al., Generating haikus from images (2021). https://medium.com/geekculture/generating-haikus-from-images-c0d35c2470ce
21. R. Rzepka, K. Araki, Haiku generator that reads blogs and illustrates them with sounds and images, in *Proceedings of the Twenty-Fourth International Joint Conference on Artificial Intelligence* (2015)
22. https://en.wikipedia.org/wiki/Internet_of_things
23. https://en.wikipedia.org/wiki/Genetic_algorithm
24. E. Gurer, A computational approach to generate new modes of spatiality. A|Z ITU J. Fac. Archit. (2016)
25. Prabandh, Vastu and Rupak. http://hindustaniclassicalmusic.in/article1.php?sub_type_id=87
26. Ashok Ranade Interviews Pt Mallikarjun Mansur (2011). https://www.youtube.com/watch?v=pMxP3O3mxA8
27. W. Claire, *Khayal: A Study in Hindustani Vocal Music*, vol. I (University Microfilms International, Ann Arbor, Michigan, USA, 1971)
28. P. Atre, Suswaravali, Bookmark Publication (2011)
29. https://www.youtube.com/watch?v=iO6_Sb3Amaw. Payaliya Jhankaar—Ajay and Abhijit
30. Shirish Pai and Haiku, Samvaad, Editorial@Maharashtra Times, Sep 24 (2017). https://maharashtratimes.indiatimes.com/editorial/samwad/shirish-pai-and-haiku/articleshow/60807211.cms
31. https://en.wikipedia.org/wiki/Doha_(poetry)
32. https://en.wikipedia.org/wiki/Kabir
33. http://www.tanarang.com/
34. https://ragakosh.com/
35. Vasant, Sangit Visharad, Sangeet Karyalaya, Hathras
36. https://www.aathavanitli-gani.com/index.htma
37. : https://en.wikipedia.org/wiki/Regression_analysis
38. https://en.wikipedia.org/wiki/Logistic_regression
39. https://builtin.com/data-science/step-step-explanation-principal-component-analysis
40. https://en.wikipedia.org/wiki/Eigenvalues_and_eigenvectors
41. https://en.wikipedia.org/wiki/K-means_clustering
42. A. Salgaonkar, M. Deshpande, S. Bodhankar, Tipedi: Marathi Haiku Bee, Pradnyarth Publication (2018)
43. https://www.youtube.com/playlist?list=PL-WsoyY1ErChT3N2mBF724yvjop5_is75
44. Swar-Govinda, Marathi Bandish concert. https://www.youtube.com/watch?v=dAXG8-rUeFg
45. https://en.wikipedia.org/wiki/Ensemble_learning

Composition and Choreography

Composing Music by Machine Using Particle Swarm Optimization

Siby Abraham and Subodh Deolekar

Introduction

Composing music requires theoretical knowledge and years of music training. A composer goes through various stages from the day he/ she starts learning music. One first understands the basic theory, which consists of music notations and mnemonics. Once the learner masters the mnemonics, he/she tries to learn some predefined compositions. Then the next step is to create a new composition based on the knowledge gathered. The learner uses feedback of himself/ herself and others to edit and modify the composition made. Gradually, he/she becomes an expert composer.

The following successive steps can describe the whole process of music composition:

- Getting an inspiration/idea or a clue from another composition
- Planning a composition
- Creating a rough composition
- Critical assessment of the composition
- Editing the composition
- Final composition

Though there are different steps involved in music composition, the process can be very brief or lengthy. For instance, the legendary Mozart has composed three

S. Abraham
School of Business Management, SVKM's NMIMS Deemed to be University, Mumbai 56, India
e-mail: siby.abraham@nmims.edu

Department of Computer Science, University of Mumbai, Mumbai 98, India

S. Deolekar (✉)
Department of Computer Science, University of Mumbai, Mumbai 98, India
e-mail: subodhdeolekar@gmail.com

A. Salgaonkar and M. Velankar (eds.), *Computer Assisted Music and Dramatics*,
Advances in Intelligent Systems and Computing 1444,
https://doi.org/10.1007/978-981-99-0887-5_9

symphonies in a single summer in the late 1970s. However, Beethoven used to spend years on a single theme.

Irrespective of the composition's duration, a composition is considered melodic when it is the outcome of diversification and intensification. Diversification is the process of searching through many different/broad styles/genres. Intensification is regarded as in-depth experimentation with similar types. The apt combination of diversification and intensification brings out a refined and soothing composition.

Suppose we can develop a methodology that automatically offers diversification and intensification. In that case, music composition's creative process can be converted to a computational process approach is what we are trying to do in this chapter. The composition of strokes of tabla, an Indian percussion instrument, is used for this purpose. Particle swarm optimization (PSO) [5], a machine learning technique, is used to generate aesthetically pleasing and grammatically correct compositions. There are many algorithm-based approaches to Western music composition [8, 9, 11], [6, 14]. However, only a few of them focus on western music, and hardly any of them concentrate on PSO.

The chapter is divided into six sections. Section II gives an overview of tabla as a musical instrument and the composition of music in tabla. Section III introduces particle swarm optimization (PSO) as a machine learning algorithm. Section IV deals with the method proposed, which formulates a strategy to convert a candidate music composition into a particle and apply PSO. Section V discusses the results and findings based on the experiments conducted. Section VI presents the conclusion and directions for future work.

Composing Music Using Tabla

Tabla is one of the recognized and highly played percussion instruments of Indian classical music [1]. It consists of two components-a right side drum, called dāyān, and a left side one, known as bāyān [13]. A typical tabla set consisting of these is shown in Fig. 1.

The strokes which are played on the table are known as bōls. A list of basic bōls played on dāyān, bāyān and both together is shown in Table 1. These syllables are onomatopoeia (words that sound like what they represent or connote). They vary according to different schools of tabla [10].

Tabla produces sound through being struck in a sequence, and the process generates rhythm, called tāl. Figure 2 shows the concept of rhythm and timing details of a common tāl in tabla called 'jhaptāl. ' Here 'mātrā' is a single beat. The rhythmic cycles or structures (called as tāls) are fixed patterns of similar length made of 'mātrās.' They are split into sections called 'vibhāg. ' The whole complete cycle is called an 'āvartan.' Tabla players provide a regular check-point in the form of the first beat of the cycle in a tāl, called 'sam' for keeping rhythmic accuracy which is used to set the pace of music [2].

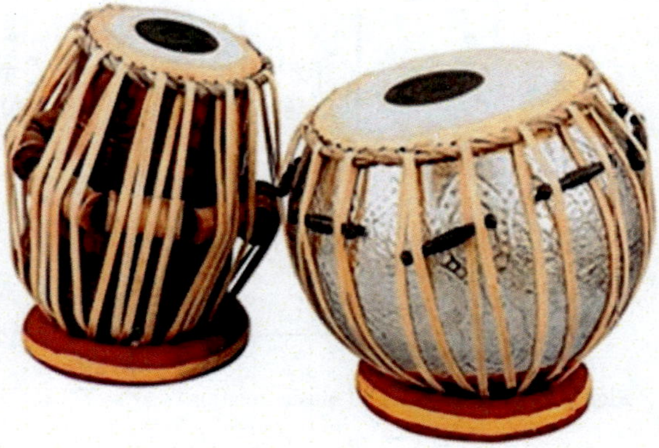

Fig. 1 Tabla set

Table 1 List of basic tabla bōls

Daayaan Bols	***Baayaan Bols***	***Both Together***
Ge (गे/ घे)	Na (ना)	Dha (धा)
Ka (क/ की)	Tu (तू)	Dhin (धिं)
	Ti (ति)	Tin (तिं)
	Ta (ट)	
	N (न)	
	T (त/ र)	
	Tra (त्र)	
	Din (दिं)	

When we take a closer analytical look at composing rhythmic music, we observe some definite principles that define the process.

The music composer responsible for creating compositions must have a command over a considerable amount of knowledge-base with various bōls. While creating a new composition, he must possess a deep understanding of the compositions' rules

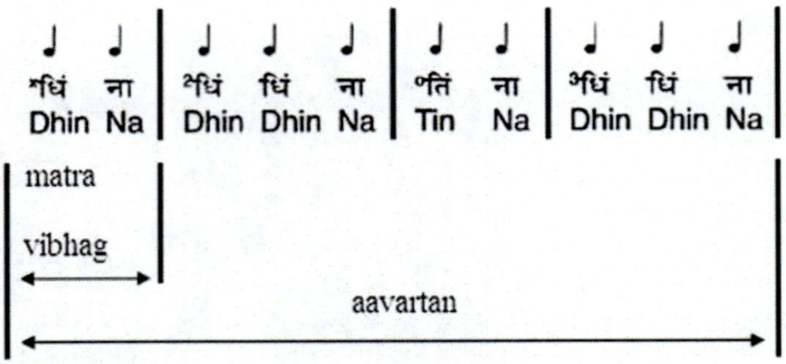

Fig. 2 Jhaptāl (Hindustani music notation)

and aesthetics. With this knowledge-base, an attempt for creating new compositions and their improvisation is made. He uses his vast knowledge gathered over long years and experiments with new improvisations making it more creative.

These processes can be automated by generating a vast collection of candidate compositions, which follow the tabla rules and grammar. A better composition can be identified from this enormous collection of compositions using machine-driven rules, which take care of the compositions' technical and aesthetic sensibilities. If we can do these processes iteratively over a considerable number of candidate solutions, we can finally get a better composition which we are aiming at. We are attempting it in this paper using one machine learning algorithm called particle swarm optimization.

Particle Swarm Optimization (PSO)

Particle Swarm Optimization (PSO) is an optimization technique that iteratively tries to improve potential solutions and searches for an optimal solution by updating candidate solutions' status over generations [5]. PSO is inspired by the social behavior of bird flocking or fish schooling. Candidate solutions, known as particles, move around the search space by following the current optimum particles' position and velocity. PSO starts with the initialization of a population of random solutions. Every particle's movement is influenced by its local best-known position at each iteration and is guided towards the global best positions in the search space [12].

The particles' movement is guided by its current position, given in Eq. (2), and the velocity of the particle at that position, given by Eq. (2). The different steps involved in the movement of the particles based on these values are given in Fig. 3. The figure shows that a typical PSO is initialized by offering random positions to a specified number of particles. Then, each of them gets updated with two values-pbest, the personal best of each of the particles, and gbest, the best of all particles till then. Once these best values are defined, the particle updates its velocity and position defined by the following equations:

Fig. 3 Flowgraph of particle swarm optimization technique

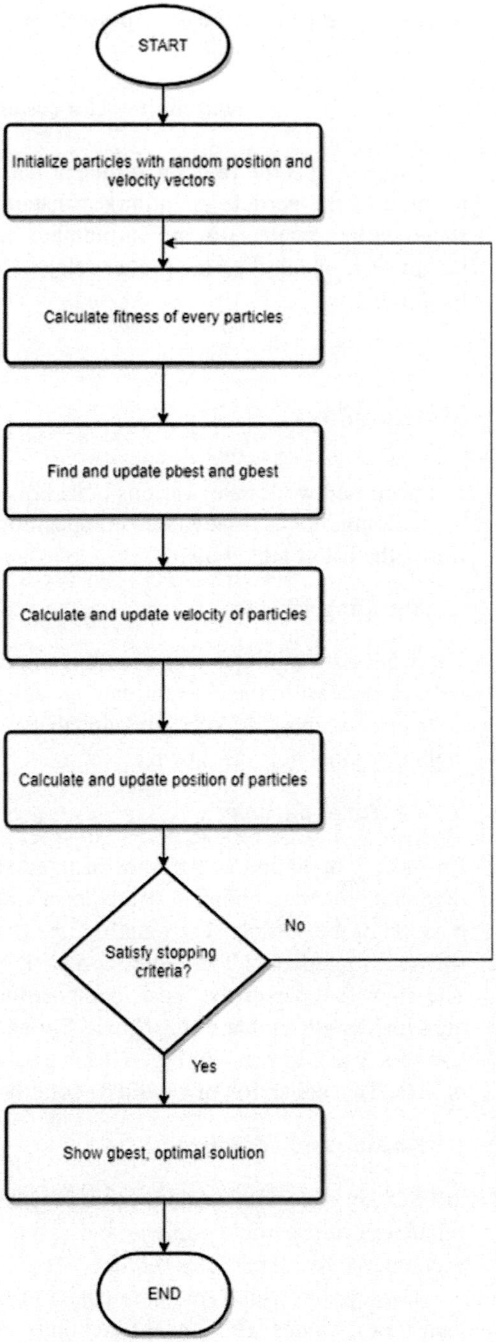

$$\text{vel [t + 1] = vel [t] + s1 * rand() * (pbset[t] - current[t]) + s2 * rand() * (gbest[t]) - current[t])} \tag{1}$$

$$\text{current [t + 1] = current [t] + vel [t + 1]} \tag{2}$$

Here, vel [t] is the velocity of the particle at instant t, and current[t] is the current position of the particle (solution) at instant t. pbest[t] and gbest[t] are defined as stated earlier. rand() is a random number between 0 and 1. s1 and s2 are learning parameters, which the practitioner selects to control the behavior and efficiency of the PSO.

Methodology

The proposed work takes various tabla bōls as input in textual format. The 32 bōls, the building blocks of any table composition [2], are taken as the input set. Table 2 shows the list of tabla bōls.

a. **Encoding scheme**

The work uses a unique alphanumeric encoding scheme to represent various tabla bōls, in contrast to the conventional bit, integer, natural, or permutation representations [4]. Numbers' 0' to '9' and alphabets 'A' to 'V' are used as substitutions of the tabla's original bōls, as shown in Table 3.

b. **Swarm of particles**

An initial population of particles of fixed size and their positions is formed using randomly generated bōls. A string consisting of bōls of fixed length is taken as the position of the particle. The length of the string depends on the type of composition we want to generate. The 'goal' composition is assumed to be known in advance. The algorithm parses the 'goal' composition and randomly generates variations of the same length as that of the 'goal.' For example, if the goal composition is of the size of say, 10, as given in Fig. 4, each particle will be a string of length ten formed of bōls. This collection of candidate solutions is taken as the swarm of particles.

c. **Position and velocity**

In PSO, we start with random positions and velocity vectors for each particle. We calculate every particle's fitness and update its best position and the best position encountered by all particles till then.

The particles, which are nothing but initial random compositions, are checked for their fitness values. The fitness function is implemented using the concept of 'fuzzy string matching. It calculates which string is closest to the goal using Levenshtein distances [15]. The Levenshtein distance is a numerical measure of the number of steps needed to replace the original character with the desired one. The Levenshtein

Table 2 List of bōls

Right-hand bōls

ता	तिं	तू	ति	ति्	ते	त	ट	त्र	र	र	ना	न	दीं	ड़	डा
					त्										
T	T	T	T	T	Te	Ta	Ṭa	Tr	Ra	R	N	N	D	D	D
a	i	u	i	i	t			a		e	a	e	i	a	a
a	n										a		n		

Left-hand bōls

गे	गि	ग	घे	घि	घ	के	की	क	क
									त्
G	G	Ga	Gh	Ghi	G	Ke	Ki	K	K
e	i		e		ha			a	at

Both together

धा	धिं	धिं	धि	धि्	धे
					त्
Dh	Dh	Dh	D	Dh	D
a	in	un	hi	e	he
					t

distance between two strings' a' and 'b' can be calculated mathematically by the function **lev**$_{a,b}$ **(|a|, |b|)**, which is given by Eq. (3):

$$lev_{a,b}(i,j) = \begin{cases} \max(i,j) if \min(i,j) = 0, \\ \min\begin{cases} lev_{a,b}(i-1,j)+1 \\ lev_{a,b}(i,j-1)+1 \\ lev_{a,b}(i-1,j-1)+1_{(a_i \neq b_j)} \end{cases} & otherwise. \end{cases} \tag{3}$$

Here, 'i' and 'j' are the initial characters of strings' a' and 'b,' respectively. The lesser the distance between the 'goal' composition and the given composition, the larger its fitness value.

We generate the lth particle with random position θ_l in the range $[\theta_{\max}, \theta_{\min}]$. The initial velocity for the lth particle $\upsilon_l = 0$. We calculate the personal best' pbest' $\psi_l = \theta_l$ for $l = 1, 2, \ldots, L$.

Table 3 Alphanumeric representation scheme to represent tabla bōls

Right-hand bōls

ता	तिं	तू	ति	ति्	ते त्	त	ट	त्र	र	र	ना	न	दीं	ड़	डा
T a a	T in	T u	T i	T i	Te t	Ta	Ṭa	Tr a	Ra	R e	N a a	N e	D i n	D a	D a
0	**1**	**2**	**3**	**4**	**5**	**6**	**7**	**8**	**9**	**A**	**B**	**C**	**D**	**E**	**F**

Left-hand bōls

गे	गि	ग	घे	घि	घ	के	की	क	क त्
G e	G i	Ga	Gh e	Ghi	G ha	Ke	Ki	K a	K at
G	**H**	**I**	**J**	**K**	**L**	**M**	**N**	**O**	**P**

Both together

धा	धिं	धिं	धि	धि्	धे त्
Dh a	Dh in	Dh un	D hi	Dh e	D he t
Q	**R**	**S**	**T**	**U**	**V**

Dhin	Na	Dhin	Dhin	Na	Tin	Na	Dhin	Dhin	Na

Fig. 4 Goal composition (tāl jhaptāl) of length 10

We update the pbest for $l = 1, 2,\ldots,$ L with $\psi_l = \theta_l$, if $f(\theta_l) < f(\psi_l)$, where $f()$ is the fitness value. The fitness function is implemented using the concept of 'fuzzy string matching. It calculates which string is closest to the goal using Levenshtein distances (refer to Sect. 5.1.3.1.2). We also update the global best of all particles till, denoted by 'gbest' for $l = 1, 2,\ldots,$ L with $\psi_g = \psi_l$, if $f(\psi_l) < f(\psi_g)$.

We then update the velocity and position of a particle using the respective equation described earlier.

d. **Termination condition**

The algorithm is terminated when the composition, which is almost matching with the goal composition, is provided by the PSO as decided by the fitness function. To reduce the computational time, which might take to look for an elusive best fit composition matching with the goal composition, provision for a maximum number of iterations is also provided. In this way, in the unforeseen situation of the algorithm not finding a good composition, it will halt after trying for a fixed number of iterations.

Results and Discussion

We have used the Python programming language to implement our scheme. The output is generated in the form of MIDI (Musical Instrument Digital Interface) for the evaluation. We have used JavaScript to generate MIDI output. To evaluate the performance of PSO. We have run several tests with sample data as various tabla compositions. The model assumes a goal composition as input and randomly creates a population of particles using the bols of the table. It does not need any other parameter settings.

a. **Fitness evolution**

Figures 5, 6, and 7 show the change rate of fitness values concerning the generation number. We have taken generation numbers along the X-axis and average fitness values along the Y-axis. Different tāls of lengths 10, 12, and 16 were used with the number of particles as 100.

It has been observed that the average fitness value of a population of particles stays the same or gets better from iteration to iteration. But it took a significantly large number of iterations to reach the goal solution.

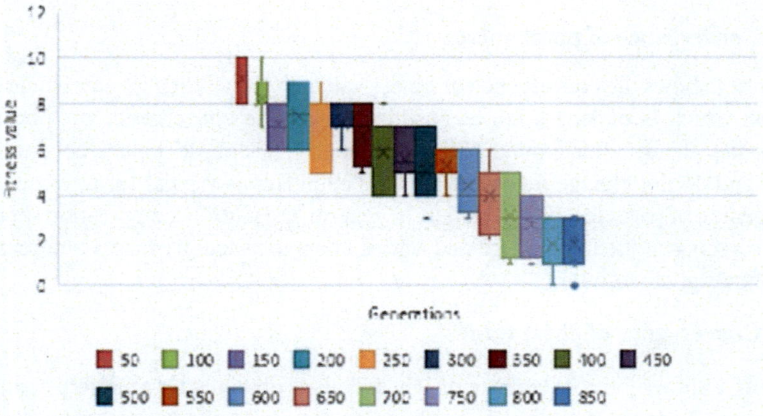

Fig. 5 Evolution of fitness values for the composition of length 10

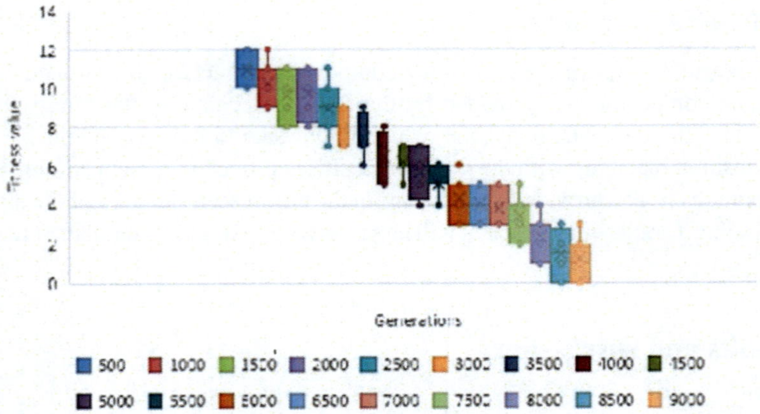

Fig. 6 Evolution of fitness values for the composition of length 12

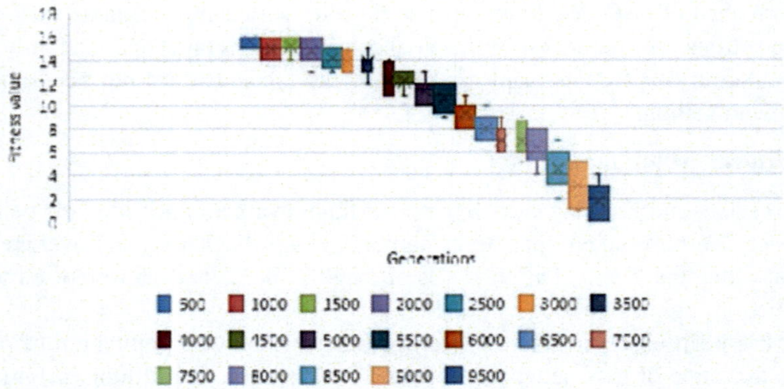

Fig. 7 Evolution of fitness values for the composition of length 16

b. **Convergence of pbest values**

Figure 8 shows the convergence of fitness value is 'pbest' for a given particle composition which is of length ten as an illustration. The convergence of 'pbest' value becomes steady after around 500 generations. The 'pbest' values become consistent and do not change much towards the end. This is typical for any convergence procedure of particles' pbest' values. In general, the particle composition shows slow convergence at the initial stage and settles down to a steady process before reaching the final 'goal' solution.

c. **Convergence of gbest values**

Figure 9 shows the convergence of 'gbest' values obtained by a particle composition of length 10 in the population. As shown in the figure, the convergence is suddenly faster at the initial stages than the convergence of 'pbest' values. Since each particle's

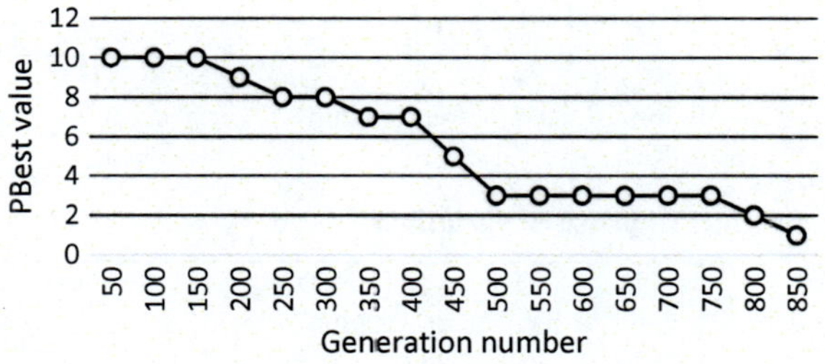

Fig. 8 Convergence of pbest values across generations

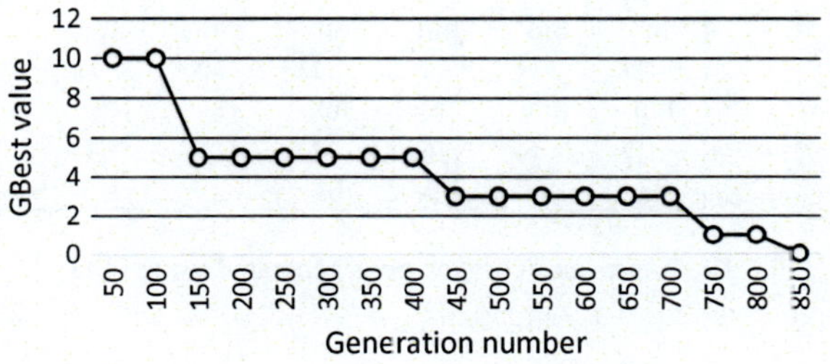

Fig. 9 Convergence of Gbest values across generations

fitness value at the initial phase of convergence is much faster, we observe sudden changes in the 'gbest' values. Towards the end, the process becomes mature and steady, resulting in a slow rate of convergence.

d. **Improvisation of the original tāl**

We have considered jhaptāl to demonstrate the improvisation process by PSO. Figure 10 shows a variety of different compositions generated at the successful completion of PSO for the input tāl. All these are valuable compositions.

e. **Evolution of improvisation**

Figures 11, 12, and 13 show the improvisation process's evolution for different numbers of particles: 20, 50, and 100, respectively. The number of iterations required is less when the number of particles is less as the computations involved are more minor. As the population size of the particles increases, the number of iterations required also increases correspondingly. There is a lot of fluctuation in the fitness function values initially. However, as the process becomes matured, there is a gradual convergence to the local optimum composition.

f. **Overhead incurred in the improvisation process**

Figure 14 shows the computational cost incurred with the different particles to reach the final composition with different lengths. The three compositions considered were of lengths 10, 12, and 16, respectively. The number of particles chosen was 20, 50, and 100, respectively. It may be noted that a new and valid composition was generated in a little over one minute, starting with all three input compositions and 20 particles.

Original tāl

D	N	D	D	N	Ti	N	D	D	N
hin	a	hin	hin	a	n	a	hin	hin	a

Improvised variations

D	D	N	D	N	Ti	N	D	D	N
hin	ha	a	hin	a	n	a	ha	hin	a
D	N	D	D	N	T	N	D	D	N
ha	a	ha	ha	a	u	a	hin	hin	a
D	D	D	D	N	Ti	N	D	D	N
ha	hin	hin	hin	a	n	a	ha	ha	a
D	D	N	D	N	Ti	D	D	D	N
hin	hin	a	hin	a	n	ha	hin	ha	a
D	N	D	D	N	T	N	D	D	N
ha	a	hin	ha	a	u	a	ha	ha	a

Fig. 10 Improvised compositions generated using PSO

Fig. 11 Convergence of PSO process with 20 particles

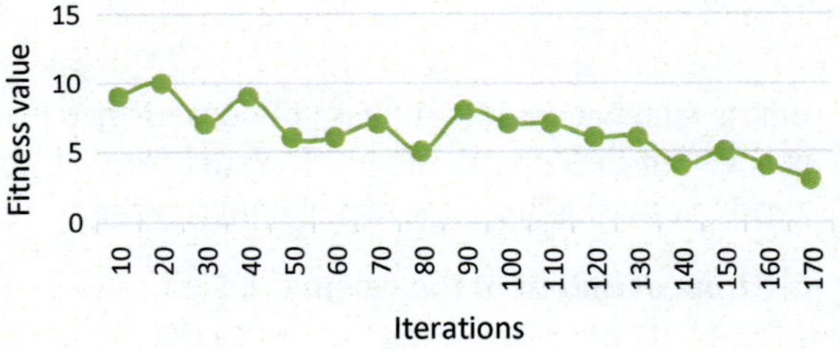

Fig. 12 Convergence of PSO process with 50 particles

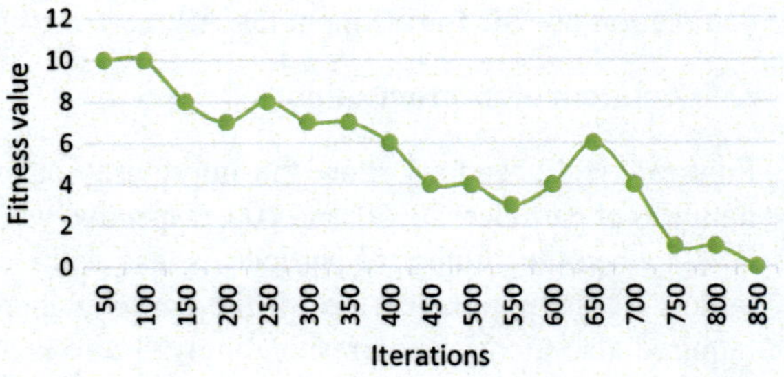

g. **Validation of improvisation process**

We have arranged a human listening test with music experts to observe whether the final compositions generated using PSO improved from the initial population of compositions or not. Five music experts, having a degree equivalent to a bachelor's degree in tabla and around 10–15 years of professional experience in the field, were commissioned for the purpose.

Fig. 13 Convergence of PSO process with 100 particles

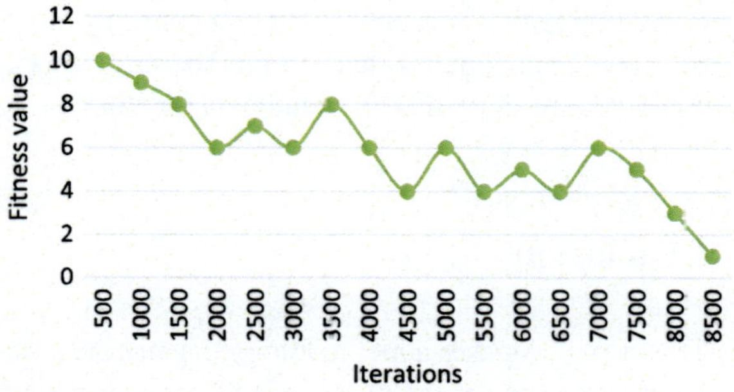

Fig. 14 CPU Time (In Seconds) and number of particles

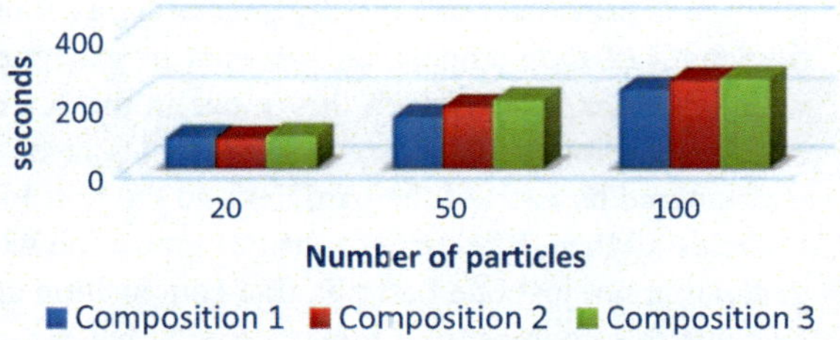

A pair of compositions were selected for the test so that the first composition was chosen from the initial population of random formation of composition. In contrast, the second one was chosen from the last iteration of the compositions provided by PSO. The music experts were asked to listen to the first one and then the second one immediately after the first. They were also asked to give a score on a scale of 1 to 10, 1 being the worst composition and 10 being the best one. The results obtained from the experiment are presented in Table 4.

We observed that those optimized compositions as the improvised composition by PSO were ranked well, as given by the average score in the table, by the music experts. We compared the scores obtained from different experts, and we could see that the experts evaluated compositions almost (between the range of 7 to 9 in the table) in a similar manner. The average score obtained for the composition in the

Table 4 Result of human listening test

	EXPERT #	1		2		3		4		5	
COMPOSITION#	Composition Pair	1	2	1	2	1	2	1	2	1	2
	1	1	9	1	9	1	9	1	9	1	7
	2	1	8	1	8	1	7	1	9	1	6
	3	1	8	1	8	1	8	1	9	1	8
	4	1	6	1	7	1	6	1	7	1	5
	5	1	8	1	9	1	7	1	9	1	7
	Average score	**1**	**8**	**1**	**8**	**1**	**7**	**1**	**9**	**1**	**7**

initial population was one, and it had gone beyond 7 for the compositions generated by PSO. This conveys that the methodology proposed using PSO could generate compositions classified as suitable by the domain experts.

Conclusion

This paper proposes a methodology to generate tabla compositions by machine. The algorithm used for the purpose is particle swarm optimization, a machine learning algorithm. It starts with a fixed number of agents called particles. The particles' position is provided randomly by generating a set number of bōls, depending upon the length of the composition we want to generate. These particles are provided with velocities, using which they move in the search space. The locations (which are compositions) chosen by these particles next are driven by the best location encountered by each of the particles and the best location, which is seen by all the particles till then. Effectively, music composition works as an optimization process, generating not just one best possible composition at the end but a collection of the best possible compositions looking similar but distinct from the goal composition.

The proposed method may have its commercial possibilities too. Based on a suggestion or input, the machine can offer possible compositions based on the same genre. It provides the opportunity for automatic music composition with minimal human intervention on a large scale, thereby augmenting a human composer's efforts to a considerable extent.

References

1. D. Courtney, The Cadenza in North Indian Tabla. Percussive Notes, Lawton OK **32**(4), 54–64 (1994)
2. D. Courtney, Learning the Tabla. Volume 2, M. Bay Publications (2001)
3. S. Deolekar, S. Abraham, Tree-based classification of tabla strokes. Curr. Sci. **115**(9), 1724–1731 (2018)
4. S. Deolekar, N. Godambe, S. Abraham, (2018). Genetic algorithm to generate music compositions: a case study with table. In: A Abraham, P Muhuri, A. Muda, N. Gandhi (eds.) hybrid intelligent systems. HIS 2017. Adv. Intell. Syst. Comput., **734**. Springer, pp. 331 -340
5. R. Eberhart, J. Kennedy, A new optimizer using particle swarm theory, in *Proceedings of the Sixth International Symposium on Micro Machine and Human Science, Nagoya, Japan, Piscataway, IEEE Service Center, NJ,* (1995) pp.39–43.
6. J. Fernandez, F. Vico, AI methods in algorithmic composition: A comprehensive survey. J. Artif. Intell. Res. **48**, 513–582 (2013)
7. J.Kennedy, R. Eberhart, A discrete binary version of the particle swarm algorithm, in *Proceedings of the Conference on systems, Man, and Cybernetics, IEEE Service Center, Piscataway, NJ,* (1997), pp.4104–4109
8. J. Maurer, (1999) A brief history of algorithmic composition. Unpublished manuscript. Retrieved from https://ccrma.stanford.edu/~blackrse/algorithm.html
9. G. Nierhaus, *Algorithmic composition: paradigms of automated music generation* (Springer, Berlin, Heidelberg, 2009)

10. A. Patel, J. Iversen, Acoustic and perceptual comparison of speech and drum sounds in the north indian tabla tradition: an empirical study of sound symbolism, in *Proceedings of the 15th International Congress of Phonetic Sciences (ICPhS)*, Barcelona, Spain (2003)
11. M. Pearce, G. Wiggins, Towards a framework for evaluating machine compositions, Proceedings of the Symposium on Artificial Intelligence and Creativity in Arts and Science, (2001) pp. 22–32
12. D. Rini, S. Shamsuddin, S. Yuhaniz, Particle swarm optimization: technique, system, and challenges. Int J Comput Appl **14**(1), 19–27 (2011)
13. S. Saxena, The art of tabla rhythm: essentials, tradition, and creativity. Sangeet Natak Akademy & D.K. Printworld (P) Ltd (2006)
14. M. Simoni, R. Dannenberg, Algorithmic composition: a guide to composing music with nyquist. University of Michigan Press (2013)
15. S. Zhang, Y. Hu, G. Bian, Research on string similarity algorithm based on Levenshtein Distance, in *Proceedings of IEEE 2nd Advanced Information Technology, Electronic and Automation Control Conference (IAEAC)*, Chongqing, (2017) pp. 2247–2251

Computable Aesthetics for Dance

Sangeeta Chakrabarty and Ramprasad S. Joshi

Introduction

It is argued, in expository literature, that BharataNatyam consists of *movements conceived in space mostly either along straight lines or triangles* (*A Web Dictionary of Indian Classical Dances: Bharatanatyam*). Vatsyayan [14] says, "the dance composition a highly elaborate edifice on the foundations of …repetitive melody" and the poses that the dancer strikes are "sculpturesque," partly because "she holds a stance in a given point of time" and partly because of the need to realize the moods and attributes of her characters as laid down in iconography. Moreover, her "treatment of space is similar to the sculptor's and each single unit of movement of the human form is significant in so far as it is related to the ultimate objective of evoking the particular emotive state (Rasa)".

Similarly, Isadora Duncan, one of the most famous danseuses of all times, who both imbibed and broke traditions and theorized and founded new ones, says (cited in [3]) that the source of dance is "Nature."

The examples cited above indicate the natural belief that universals of aesthetics must be sought in Nature. We review three sources here: Biology, Neuroscience, and Psychology. We posit that the universals will be retained, across space and time, regardless of the changes and mutations naturally occurring in the original traditions during the long evolution.

Sangeeta Chakrabarty is a.k.a Sangeeta Jadhav.

S. Chakrabarty (✉)
Department of IT, S. S. Dempo College of Commerce and Economics, Cujira, Goa, India
e-mail: sangeeta.chakrabarty@dempocollege.edu.in

R. S. Joshi
CS-IS Dept K K Birla Goa Campus, Birla Institute of Technology and Science, Pilani, Zuarinagar, Goa, India
e-mail: rsj@goa.bits-pilani.ac.in

A. Salgaonkar and M. Velankar (eds.), *Computer Assisted Music and Dramatics*, Advances in Intelligent Systems and Computing 1444,
https://doi.org/10.1007/978-981-99-0887-5_10

What is retained recognizably, as universals, is also expected to be amenable to computational modeling. Mallik et al. [8] precisely argue this in their work proposing an architectural framework that includes a method to construct the ontology with a labeled set of training data and use of the ontology to automatically annotate new instances of digital heritage artifacts depicting Indian classical dance forms: "Once a multimedia enriched ontology is available, it can be used to interpret the media features extracted from a larger collection of videos, to classify the video segments in different semantic groups, and to generate semantic annotations for them. The annotations enable creation of a semantic navigation environment ..."

Automation for dance also has two potential implementations: classical symbolic computation using generative grammars and data-driven machine learning. Grammars are condensed, encoded models of cumulative collective wisdom. Machine-learning models also represent the contribution of evolution through generations of practitioners. Machine-learning can be statistical, relational, inductive logical, and connectionist (neural networks). Both representations are partial and never complete descriptions of the ever advancing actual tradition and practice.

Symbolic computation relies upon the condensed codification usually with scholarly recognition of classical traditions. In the interest of universality, that is, to avoid cultural fixation, we must put together this codification with data-driven models that express the universals in practice more than the codified, recognized tenets. This needs exhaustive data, and that is the challenge. Not just classical dance forms, but we do not have good enough datasets for classical languages. We cannot just hope to fill the gap by any other means to apply supervised learning fruitfully in this case. We must embed reinforcement learning processes into the practice and craft of artistic performances. In this work, we consider this challenge as the central concern. And we argue that recognition of the existence of universals beyond their actual codification in grammar should help us meet this challenge.

Previous Work

The idea of computing for aesthetics is not new, though recent. The researchers try to locate the possibilities again in perceived natural universals.

Biology, Neurology

Machado and Cardoso [7], an early attempt at "computing aesthetics," argue that "visual aesthetic value, is connected to visual image perception and processing, thus being mainly: biological, hardwired, and therefore universal. We aren't saying that it is coded in our genes, we just mean that what we are, and the way our visual image perception system works, makes us favor certain images to others."

Learning from Data

An M.Tech. thesis [11] clustered a large set of BharataNatyam poses to obtain a *bag of posewords* to model these dance sequences, then used a histogram of posewords as a feature vector for classification using standard learning techniques (in this case, SVM) and also built a Hidden Markov Model (HMM) to recognize basic dance steps of BharataNatyam (an Indian Classical Dance) using RGB + depth (RGB–D) video data obtained from MS Kinect. The accuracies obtained are 68% (SVM) and 81% (HMM). Finally their action recognizer used voting scheme on multiple HMMs to build a robust action classifier with a cross validation accuracy of 100% on their dataset of 120 BharataNatyam videos.

Learning the Grammar

Both the above approaches assume universals, which makes us wonder: is it possible to learn the grammar of dance (or of visual art and music also) from data? By grammar, here we mean the way compositions are structured, with the lexicon having, named or unnamed, basic poses, stances, gestures, steps, and movements that are recognized as elements of the compositions. In classical cases, like BharataNatyam, the lexicon is codified well and the grammar also specified in as much explicit form as is possible in performance arts.

Similar motivation led Nakazawa and Paezold-Ruehl [9] to build a "proof of concept system that uses genetic algorithms to generate choreography for the waltz, a ballroom dance." However, according to them, waltz "steps are designed so that one can generally link them into any other step, producing exponentially large numbers of combinations. Due to the physical limitations of the performance space, generating choreography is also a constraint based optimization problem. Both of these attributes suggest that genetic algorithms are suited to finding a solution." Thus, they relegate the aesthetics question to constrained optimization and reduce learning to local search. We view this reduction too restrictive and pessimistic. Aesthetics is neither so simple as to be reducible to numerical optimization, nor is it so complex to be beyond computation.

The Flow: The Learning and Automation Loop

The present work is a natural consequence of the first author's thesis [1]. This work sought to automate generation of choreographic pure dance movement sequences befitting the BharataNatyam framework by modeling the formalized BharataNatyam pure dance movements (traditionally called *adavus*) by a grammar on limb movements.

This is essentially the same (but generative) approach as that of the masters thesis [11] which is in the opposite, cognitive direction.

In Chakrabarty [1], aesthetic acceptance of *grammatically* generated sequences proposed or classified by soft computing techniques was taken from human experts. We argue that it would help to assume existence of universals of aesthetics in order to build a closed-loop system composing the complementary approaches of the two theses.

The sections in the rest of the paper are organized as follows. We first discuss Ramachandran's identification of neurological universals of aesthetics. Then we discuss their relevance to BharataNatyam and Chakrabarty's work [1]. Further, we discuss the current difficulties in getting datasets vetted by domain experts for advancing work in the direction of [1]. Subsequently, we draw from a deep probability inequality a surprising conclusion about why it should be possible to ease these difficulties using video and image analysis by machine learning equipped with condensed domain knowledge in ontologies. At the end, we sketch an outline of such an approach.

Universals from Neuroscience: Ramachandran's "Navarasas" or 9 Aesthetic Sources

Vilayanur S. Ramachandran [10, Chapter 7, "Beauty and the Brain: The Emergence of Aesthetics"] endeavors to "speculate intelligently on the neural basis of art and maybe begin to construct a scientific theory of artistic experience."

In that work, Ramachandran formulates and describes his identification of nine neurological sources of aesthetic pleasure (and correspondingly of aesthetic design and performance). Henceforth we call these nine aspects as "*navarasa*"s. While his formulations and explanations are intuitive and not computational, we seek computational modeling of the same, presently. We reinterpret his formulations with an example of our own, and alongside, we explain our ideas on how to make computational models of the same, using the same example.

Without formal and rigorous description of what it means for aesthetic ideas to be "computable," we can still talk about what is computable and what is not in relation to aesthetics. We can investigate objectively whether such ideas can improve man-machine choreography assistance system design. Before we propose ways to investigate the choreography question, we discuss computability of the navarasas in the context of one image example (Fig. 1). This picture is a photograph, taken on an 8-megapixel phone camera, of a swamp area. This is about 10×10 sq. ft., under a narrow rural road culvert, in the Western Ghats section of Maharashtra in India. At the lower end is the culvert's barrier in low-quality masonry, that is the viewpoint. Its reflection in the shallow water below is seen, as is the swampy vegetation underwater lit by penetrating evening sunrays. Some flowering is seen above the water and other vegetation forms a canopy-type background. The blue sky in the East is reflected

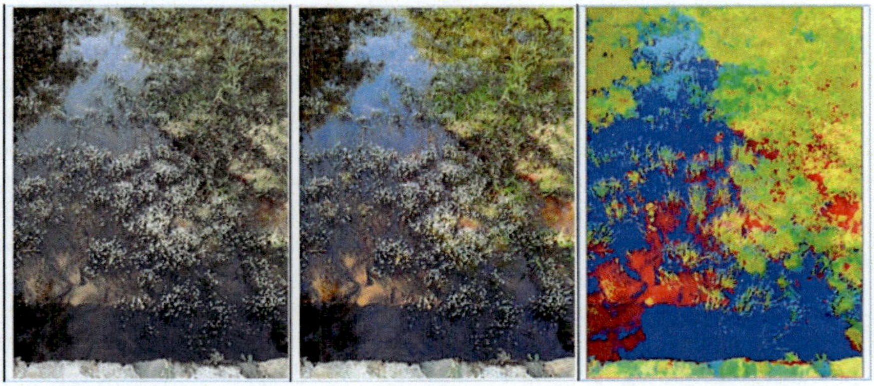

Fig. 1 L to R: contrast-reduced image of a swamp; the original image; contrast-enhanced image

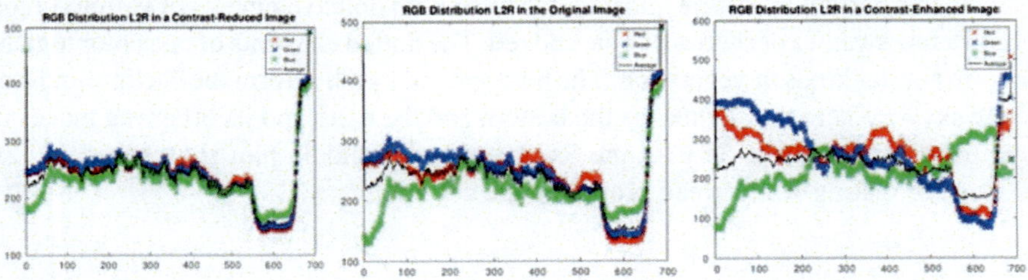

Fig. 2 Row-wise RGB sums: contrast-reduced; original; contrast-enhanced

in the water away from the culvert. Next (Fig. 2) are the plots of row-wise sums of RGB intensity values from top-to-bottom (appearing L to R in the titles) on these three pictures. The original clearly shows how the red and blue components are quite well correlated (the column-wise correlation happens to be 0.93) while the green component provides for the contrast. When this contrast is reduced, the image is lackluster; when it is too much, it is an eyesore.

We now set out to highlight the aspects of this picture relevant to the navarasas, with the corresponding computational modeling processes.

Grouping

The brain tends to group similar elements or features. *In case of BharataNatyam (BharataNatyam) successful grouping (like dressing similarly, identical looks in terms of eye makeup, jewelry, etc., height or body weight of the performers) feels good aesthetically.*

It is obvious that coming together of many similar elements renders Fig. 1 a surreal beauty. The swampiness is completely washed away by the riot of colors of life, i.e., the green of vegetation combined with the earthly color of hay and mud, the sun's rays, the sky blue, the white flowering, all share similarity of either color or texture.

Color is an easily modeled attribute, pixel to whole picture level, by simple numerical aggregates. For example, we can measure the concentration distribution of each of RGB in the image at pixel density and aggregate level to observe a quantitative indication of grouping. Our plot demonstrates just one simple way of doing so. However, the term "elements" need not be confined in its meaning as something as basic as color. Moreover, if all pixels are some shades of green, we will not see *many* similar elements coming together, but we will see *one* element dominating the picture. If the picture is full of fern then such a dominating color will appear. In that case, the variety of similar elements will be seen in the form of fractals—morphologically similar geometric elements repeating with variation in size, orientation, and combination.

In this particular picture, most of it is filled with dotted elements of various colors, and a few swathes of shades of blue and red. The dotted elements of one color together form similar kinds of vegetation. The blue and red swathes form the background from the sky—so here the sky makes the bottom and the earth and its offspring makes the superstructure and variety. In the RGB sums, the middle plot shows this: red and blue correlating while green providing the contrast.

Peak Shift

The brain responds in non-linear ways to exaggerated stimuli: e.g., why caricatures are so appealing. Essence, quintessence, *the sculpturesque, and deliberate exaggeration of the poses in BharataNatyam gives the highest peak shift. Mirror Neurons are activated while watching changing postures and movement of the body.*

Regarding Fig. 1, the swampiness that leads to a colorful variety in similarity of hue and texture actually turns into its opposite: what is otherwise ugly becomes a bed of biodiversity with beauty. Without the reflections and sun rays in the muddy water, this could not have happened. We quickly derecognize the swampiness and recognize the beauty in the colors created by the life in it and by the play of light and shadow. In a computational model, we can set aside the usual object recognition—grass and grass flowers are no more the ubiquitous spread here—and recognize pixel regions of layers of objects, split or composed together not by our perception of unity of the objects but by the perceptual unity and contrast. Thus the peak shift from the correlated red and blue (the "background sky") is provided by the greenery and flora.

Contrast

Giving variation, boundaries, spatio-temporal distinction, and movement. Too little contrast and a design can be bland. And too much contrast can be confusing. *The different colorful and contrasted costume of the background dancer with reference to the main dancer or even the accentuated eye-makeup gives a contrast.*

Had the picture in Fig. 1 been that of just swampy grass, it would be both boring and ugly. The swampiness would not go, it would be enhanced by the grass. The presence of purple water, shadows, sun rays, reflections, white flowerings, and blue sky is very much essential for the surreal beauty and suppression of the swampiness. Again techniques such as measuring concentration distribution in tessellations will give a measure of contrast. We have very sharply demonstrated this in the simple RGB sums plots.

Isolation

Emphasizing a single source of information—such as color, form, or motion—while deliberately playing down or removing other sources, e.g., a sketch can be more effective because there is an *attentional bottleneck* in the brain. *The unique hasta-mudras (hand gestures) and padabhedas (leg positions) of BharataNatyam create isolation. The main dancer as the isolated central character in BharataNatyam is dressed distinctly from the rest of the troupe, and also has different movements.*

In Fig. 1, there are many examples of isolated features highlighting an aesthetic attractor; viz., the reflection of the blue sky, sun rays under water, white flowering, etc.

Peekaboo, or Perceptual Problem Solving

It superficially resembles isolation but is really quite different: It's the fact that you can sometimes make something more attractive by making it less visible. *The BharataNatyam dancer doesn't reveal each and every detail of the story through her performance but its left to the imagination of the audience. A single gesture done by him/her could be interpreted in many different ways by every audience.*

The whole of the picture in Fig. 1 is a puzzle here, which makes one take time to recognize that it is the picture of some wild shallow stream, muddy, and swampy. The colors look sprinkled artificially until their natural belonging becomes apparent. Recognizing the "background sky" colors and imagining the time and the mood when the picture was taken is the stimulating, exciting puzzle here.

Abhorrence of Coincidences

Coincidence is suspect and unpleasant. And our brain always tries to find a plausible alternate, generic interpretation to avoid the coincidence. *The story in BharataNatyam does not progress through synchronous movements of BharataNatyam dancers, but via variations in the movements of the main dancer.*

The picture in Fig. 1 doesn't have a polluted stream, though it is a muddy one, thus retaining its natural beauty without the urban squalor we are so used to in swamps. If we look at the contrast-enhanced image, the blue and the red clearly appear artificial, but if only one of them was enhanced to contrast others, they would not appear that much artificial, and not an eyesore surely.

Orderliness

Apparently the wild, swampy scene above cannot plausibly have any order; however, the "order out of chaos" emerging in the form of abundance of symbiotic life is what makes the non-urban swamp beautiful rather than ugly. Is this more than a subjective notion, that is recognizable computationally, objectively? To be able to answer this, we need to formulate ways to express the conceptual—like symbiosis—by an operator transform, similar to spectral analysis. BharataNatyam, obviously, is all about orderliness.

Symmetry

In BharataNatyam, from the dancer's balanced dressing to the rhombus of the half-squat pose and movements symmetrically going from right to left, symmetry is everywhere. Again, it is nontrivial to see how the concentrations and distribution of the main colors in Fig. 1 are balanced out in the picture. Can this be made into a computational model? This will surely need topological-geometric tessellations on color layers.

Metaphor

In BharataNatyam the *nataraja* figure as Lord *Shiva*, the bow-armed (*kodand-dhari*) Lord *Ram*, Lord *Krishna* wearing his discoid sawtoothed weapon *sudarshan chakra* on his pinky, all are metaphorical.

Why Google's NIMA [13] works for identifying pictures with cars and dogs in them is now common wisdom, but can it classify pictures of swampy grass of

Fig. 1 and other pictures of woods separately in one sense—type of vegetation—and together in the sense of representing variety of symbiotic life forms? We are not sure, but we can infer from the NIMA experience that our metaphorical and conceptual understanding, when seeded into computational models by supervised learning, can work for narrow, singled-out aspects in an image. And, in a BharataNatyam dance performance, there cannot be too many of such identifiable metaphorical and conceptual aspects changing with movements.

Discussion: BharataNatyam, the *Navarasas* and Computation

In this section, we try to illustrate how the foregoing ideas (with some computational application to still landscape images) can be brought to be useful in computational treatment of BharataNatyam aesthetics. We show BharataNatyam poses (Fig. 3) with the relevant aesthetic principles of the nine above, and suggest their computational relevance. We have only omitted the principle of orderliness from this discussion; BharataNatyam being among the most orderly and codified classical dance form, that is a moot point.

In Fig. 3, the first 5 poses of a contemporary BharataNatyam exponent demonstrate the "new adavus" or the novel poses obtained in [4] (and approved by experts as good). The last one is the classic *Natarāja* figurine in copper (from a century old book on *AbhinayaDarpana*). Following the figures, Table 1 depicts the co-ordination and contrast between the limb positions using the 30-component vectors for each of those 5 poses thus: The HAND DIFFerence column shows the algebraic sum of the differences between the right and the left hands; the next ABSolute HAND DIFFerence column shows the sum of absolute differences of the same. Similarly, the next two columns show leg position differences.

Table 1 The table of differences

POSE	HAND DIFF	ABS HAND DIFF	LEG DIFF	ABS LEG DIFF
1	−1	9	−1	1
2	−3	17	−1	1
3	4	8	−1	1
4	0	12	−1	1
5	−7	11	−3	5

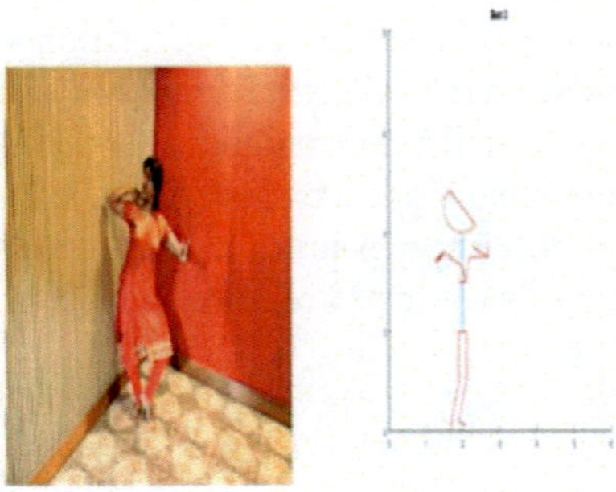

(c) actual (d) stick

21202-13-11130112100000100011000

Pose 1

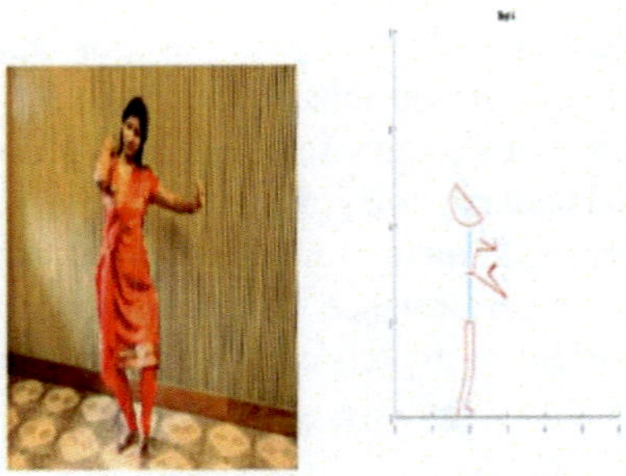

(e) actual (f) stick

21202-13-11164-2020000000100011000

Pose 2

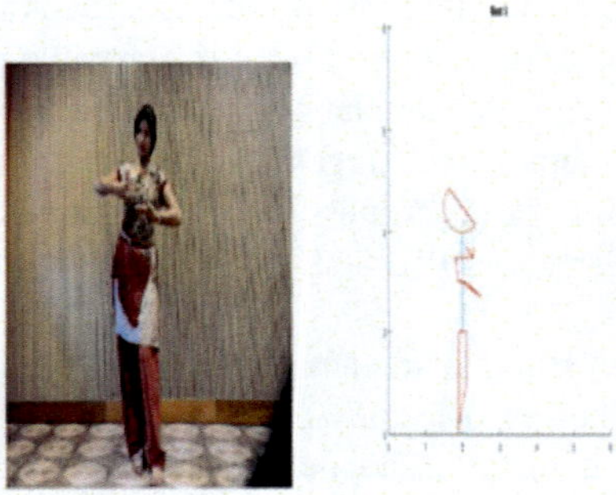

(g) actual (h) stick

212120300002-113-100000000001000

Pose 3

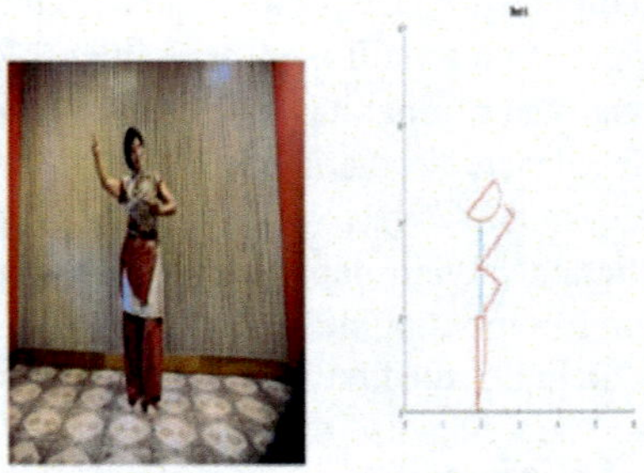

(i) actual (j) stick

1-12-220200002-113-100000000001000

Pose 4

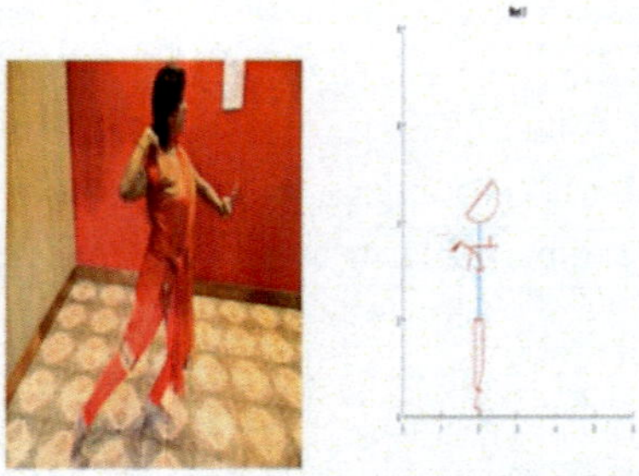

(a) actual (b) stick

1-1202-13-1-1-130112111022000-311000

Pose 5

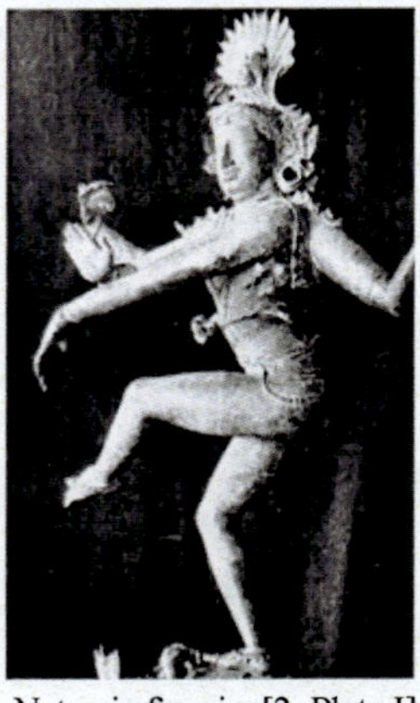

Nataraja figurine[2, Plate I]

Fig. 3 Various figures

Grouping, Contrast and Symmetry

The first two principles appear contradictory and mutually exclusive, but they are not so, as is demonstrated here. Apart from the color combinations chosen by the exponents here in the five contemporary figures in Fig. 3, the 30-component vectors and the corresponding stick figures better demonstrate the fact that similar elements are either grouped together by a co-ordinated movement or shown apart with a calibrated contrast. The movement depicts a stable stance. The poses 1, 2, and 4 show some flow, and this shows up in the component vector differences. All show low algebraic differences between hand components but the absolute differences are larger. When the same movement continues, the contrast will grow, while keeping the algebraic sum of differences small.

On the other hand, Pose 3 has the least deviation from symmetry among all the poses shown. The algebraic difference and the absolute difference are both in the same direction.

Peak Shift, Isolation, Abhorrence of Coincidences

All the three principles are exhibited in Pose 5: the danseuse is almost losing balance to one side, the center of gravity shifting, unlike the other four poses. The balance is maintained by extending both the left limbs outward. At the same time, this extension is neither antisymmetric nor coincidental. In balancing, the left leg gives support without appearing either limp or rooted. It shows isolation, which only helps focus attention. The absolute differences not being far higher than the magnitudes of the algebraic differences shows peak shift.

Perceptual Problem-Solving and Metaphor

Again, the two principles appear overlapping, if not redundant. After all, dance itself constitutes metaphorical representation of life forms. And without familiarity with the idiom and grammar of each form, just viewing cannot lead to appreciation without perception. Pose 1 involves some problem-solving because of the orientation of the head and the dancer's vision. But there is no metaphorical meaning in it. However, the Nataraja figure helps us understand the difference. The viewer is forced to wonder how the figure balances itself—though it is an inanimate sculpture or cast, it does stand like a living biped. Several other renderings of the same figure are prevalent, among relics obtained from all ages, some with a ring-frame encircling Shiva at five vertices. All show that the center of gravity does remain firmly at the center of the right (grounded) leg's toes. Performing dancers do take this pose, and staying firm and stable in it for long durations is considered one of the highest achievements of a BharataNatyam exponent. Understanding this balancing is perceptual problem-solving, forcing us to revisit our notions of stability, stance, strength. Several motifs that stand for specific symbolic meanings add to the perceptual complexity.

However, the metaphorical part is in the whole of it, wherein connecting this form, stable, yet precariously balanced stance to a turbulently fluid, Aramageddon-inducing dance. It has spawned so much literature that this point needs no more emphasis.

Discussion

This last one Nataraja is not just one of the most celebrated classic BharataNatyam poses, it illustrates most of the nine principles above. However, it is clear by now that just looking at the 30-component vectors' component groups—summing and contrasting the commonalities and differences—would not reveal much, if applied to this stance. Moreover, it invokes the three sources that prompt us first of all to ask the three questions listed as the motivation of this paper in the beginning.

The Body as an Autonomous Object

Isadora Duncan [3] says:

- The source of dance is "Nature."
- Dancing should be the natural language of the "soul."
- Dancing must express humankind's most moral, beautiful, and healthful ideals.
- Our first conception of beauty is gained from the human body.
- The Will of the individual is expressed through the dancers use of gravity.
- Movement should correspond to the form of the mover.
- Dancing must be successive, consisting of constantly evolving movements.

Vision, Perception, and Mirror Neurons

As the four pictures in Fig. 4 show, we tend to perceive performances and exhibits—ultimately human creations and craft—through the lens of our perception of the performers and creators as human beings like us. In fact, we project this anthropomorphic assumption onto other mammals, like cats and dogs; at times we are supported in these beliefs by the more intelligent animals like elephants, dolphins, and primates. Our perception of dance is mediated through our internalization of what the dancer is feeling bodily, and what she is expressing perceptually. This idea led Dr. Ramachandran, whose work on mirror neurons is one of his seminal contributions to the science of cognition, to the endeavor of outlining the neuroscience of aesthetics, the nine principles discussed above.

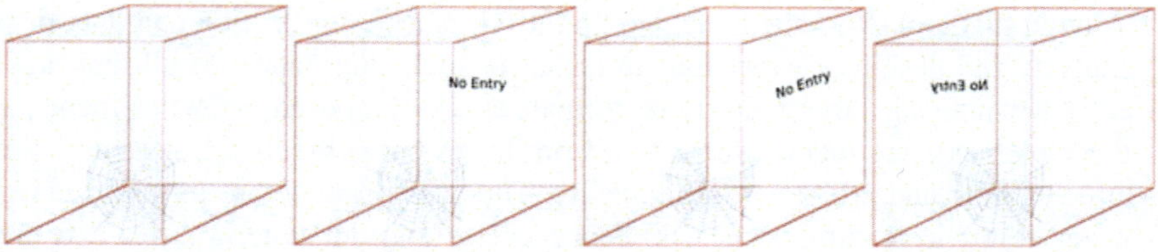

Fig. 4 Perception and Vision are not the same

We Move in 3-D, Computational Models Postulate ∞-Dimensions

We indicate one more source of optimism in that direction, apart from the powerful ideas of Ramachandran discussed above. That source is a deep result in probability theory.

In [1], Chakrabarty computed the number of *digital* possibilities of pure positions based on the degrees of freedom of each limb and component significant in characterizing dance positions was identified combinatorially. Though this number was more than 10^{18}, she argued that even basic understanding of BharataNatyam tell us that the number of plausible (and aesthetically considerable) possibilities must be a minuscule fraction of this. However, plain digital notation will not help us identify them, nor will genetic algorithms or other such soft computing that keep reliance on bit positions in the notation help much. We have to investigate the potential of making a computational model of human perception and design thinking in relation to dance, with the assumption that this must make the possibilities computationally far less numerous than the sheer combinatorial explosion in digital notation. This assumption is well justified by two sources: one, the human *attentional bottleneck* that limits the number of identified discrete items that we can pay attention to simultaneously to single digits. Two, the very sustenance and flourishing of the classical dance traditions with documentation that runs only at the most a few thousand pages, points out to the condensation of the ideas into a small number of possibilities.

Thus, the question is not really of computability of the possibilities, but rather computational complexity of their cognitive processing. In this, Talagrand's inequality [6, 12] gives us another argument in favor of its tractability. This inequality and its corresponding theory essentially tells us that in high dimensional space (like a large feature space), concentration of measure phenomenon is observed. This phenomenon ensures that large enough subsets of individual points do not differ from each other in too many dimensions.

In its most versatile and useful form, Talagrand's Inequality [12, Theorem 3.1.1] states that

> Given a set S of examples distinguished uniquely in n features and a subset A that is at least half of S, the probability that an example in S differs from any group of q examples in A in k or more features is no more than
> $$\frac{2^q}{q^k}.$$

Though this looks too technical, we find it relevant to digitization and data-driven analysis and design of every human endeavor including dance. In all their apparently bewildering variety, dance performances are composed of few elements, and those elements are themselves drawn from fewer universals that transcend boundaries of time and space across humanity. Thus, suppose we are trying to classify using aesthetic features some 100 dance videos and our criteria cover 8 features as being aesthetically influential.[1] Then, the above result says that a class of dance videos—characterized by any collection of 30-component vectors—smaller than 35 is indistinguishable from the rest on the 8 features.

In laypeople terms, if the number of apparent features and design factors is high, but the variety and diversity observed is not correspondingly widely varied in essence, then the dimensions are not really independent of each other. Models obtained by a simple generalization from our 3-D intuition to high dimensions are chosen for the ease in design and in organization of computation on them. However, this leads to an unwarranted computational intractability, as well as opacity of the essential relationships.

Conclusion

Thus we posit that Ramachandran's navarasas can be very apt targets for modeling computationally and incorporating them into an automation scheme for choreography. However, in any such endeavor, like [1], the main challenge is to get "good datasets." In order to apply ideas on aesthetics to analysis, classification or synthesis of artifacts and performances, we need to build curated datasets of good and bad artifacts and performances classified probably along multiple axes of those identified aesthetic parameters. Getting raw data (like videos, pictures, annotations, and metadata) and then getting expert opinion on various aspects is a time-consuming and costly process. The latter part, getting expert opinion, is the tougher part.

Mallik et al. [8] have noted that even equipping existing digital archives with semantic annotation "is a labor-intensive process and a major bottleneck in creating a digital heritage collection." They present a approach based on "domain ontology, enriched with multimedia data and carrying probabilistic associations between concepts" in order to "to curate a heritage collection by generating semi-automated annotation for the digital artifacts". Chakrabarty [1] took this approach and used BharataNatyam's domain ontology to create not just curated archives but use the resulting semi-automated classification for semi-automation of choreography. This was possible for BharataNatyam mainly because the ontology and the corresponding practitioner's instruction are expressed well-coded in discrete text. But is this generally applicable for other dance forms? Are Ramachandran's *navarasas* also amenable to universal computational modeling, not just for BharataNatyam?

In particular, we seek to investigate:

[1] Incidentally, it was 8 to which dimension was reduced from 30 using rough-set methods in [5].

1. Can we build video and image classification systems that use the nine criteria for aesthetic labels or evaluation?
2. Will these criteria and computational models be independent of the storytelling and audience perception? How well will it match subjective expert opinion? How to assess this?
3. If successful, can such models lead to semi-automation of choreography in general? This will of course involve
   - evolution of quasi-universal notation,
   - digitization standards,
   - automated labeling, classification, annotation, and
   - development of a computer-human interface to bring it all together into a choreography assistance system.

Future Work and Its Rationale

Both [1, 9] carried out the four steps depicted above in the third goal above of semi-automation of choreography, albeit without universal, general applicability or standard. But, then, what about the universals? Such interpretation as we made of Talagrand's Inequality and the results of [4] above gives us a two-way confidence:

1. That, the influence of aesthetic universals as source does converge into an aesthetic resonance when it comes to comprehensive design. Feature engineering from video data of human performers should converge to encoded BharataNatyam universals and Ramachandran-type universals.
2. That, it is possible to organize this computation efficiently if we remember the bounds on independence given by Talagrand's result.

References

1. S. Chakrabarty, Automation of Bharatanatyam choreography for pure dance movements. Ph.D. Thesis, Department of Computer Science & Technology (DCST), Goa University, Goa, India, 2018
2. A. Coomaraswamy, G.K. Duggirala, *The Mirror of Gesture Being the Abhinaya Darpana of Nandikesvara* (translated into English) (Harvard University Press, 1917)
3. A. Daly, *Done into Dance: Isadora Duncan in America* (Wesleyan University Press, Middletown, Connecticut, USA, 1995)
4. S. Jadhav, M. Joshi, J. Pawar, Art to SMart: an evolutionary computational model for BharataNatyam choreography, in *Proceedings of the 12th International Conference on Hybrid Intelligent Systems IEEE Xplore* (2012), pp. 384–389. 978-1-4673-5115-7
5. S. Jadhav, J. Pawar, Bharatanatyam dance classification with rough set tools, in *ICT Based Innovations* (Springer, 2018), pp. 75–81
6. M. Ledoux, M. Talagrand, *Probability in Banach Spaces* (Springer, Berlin, 1991), p. 2011

7. P. Machado, A. Cardoso, Computing aesthetics, in *Advances in Artificial Intelligence 14th Brazilian Symposium on Artificial Intelligence, SBIA'98 Porto Alegre, Brazil, November 4–6, 1998, LNAI*, vol. 1515, ed. by F.M. de Oliveira (Springer, 1998), pp. 219–228
8. A. Mallik, S. Chaudhury, H. Ghosh, Nrityakosha: preserving the intangible heritage of indian classical dance. J. Comput. Cult. Herit. **4**(3), 11:1–11:25 (2011). https://doi.org/10.1145/2069276.2069280
9. M. Nakazawa, A. Paezold-Ruehl, Dancing, dance and choreography: an intelligent nondeterministic generator, in *The Fifth Richard Tapia Celebration of Diversity in Computing Conference: Intellect, Initiatives, Insight, and Innovations, TAPIA'09* (ACM, New York, NY, USA, 2009), pp. 30–34. https://doi.org/10.1145/1565799.1565807
10. V.S. Ramachandran, *The Tell-Tale Brain: A Neuroscientist's Quest for What Makes us Human* (W. W. Norton & Company, New York London, 2011)
11. A. Sharma, Recognising Bharatanatyam dance sequences using rgb-d data. Master's thesis, IIT Kanpur, 2014
12. M. Talagrand, Concentration of measure and isoperimetric inequalities in product spaces. Publ. Math. I.H.E.S. (81), 73–203 (1995)
13. H. Talebi, P. Milanfar, NIMA: neural image assessment. arXiv:1709(05424v2) (2017)
14. K. Vatsyayan, Notes on the relationship of music and dance in India. Ethnomusicology **7**(1), 33–38 (1963)

Design and Implementation of a Computational Model for BharataNatyam Choreography

Sangeeta Chakrabarty

Introduction and Literature Review

Structure of BharataNatyam

BN is a traditional and ancient Indian Classical dance (ICD) style. Notations for BN done with stick figure drawings and *Natyashastra* (NS) and *Abhinayadarpana* (AD) are considered as the most authentic sources of Classical dance theory. This dance form has three building blocks: rhythmic dance movements (*Nritta*), representational dance (*Nritya*) and the dramaturgy (composition of both, *Natya*). The present research has been restricted to *Nritta* only. *Nritta* is composed of combinations of elementary units called *Adavus*. Limbs like hands, feet, head, etc. and facial units like eyes, move in a coordinated manner. They are established as aesthetically best for a finite number of beats comprising *Nritta*.

Problem Definition

Dance follows a rhythm and this is depicted either through only an eye movement or simultaneous movements of many body limbs. A large variety of poses and movements are possible by combining the individual movements of major limbs only, and this variety is covered in *Nritta*. Not all of the possible body limb combinations theoretically conceived are feasible (e.g. one leg in a sitting fold and the other straight for

Sangeeta Chakrabarty is a.k.a Sangeeta Jadhav.

S. Chakrabarty (✉)
Department of I.T., S. S.Dempo College of Commerce and Economics, Cujira, GOA, India
e-mail: sangeeta.chakrabarty@dempocollege.edu.in

A. Salgaonkar and M. Velankar (eds.), *Computer Assisted Music and Dramatics*, Advances in Intelligent Systems and Computing 1444,
https://doi.org/10.1007/978-981-99-0887-5_11

standing) nor are they necessarily aesthetically pleasing while maintaining the classical dance structure (e.g. both the arms straight raised up behind). Also, the movement transitions from one beat to another may have several constraints governing them.

Research effort in computation for ICD has addressed animation [1], heritage preservation [2], e-learning [3], mobile applications [4], classifying BN *mudras* [5], posture recognition for BN using Machine Learning [6], etc. But automating choreography for ICD is a recent idea, see [7]. Work on computation for Western Dances and other ICD forms than BN (till 2015) was reviewed in [8].

The rest of the chapter is organised as follows: we begin with data modelling (section "Data Modelling") followed by the description of our automation technique (section "Generation of New Dance Poses") used for designing new dance poses. Section "Generating N-Beat Dance Poses for Choreography" explains the multi-beat choreographic process, its challenges and our solutions. The final section describes the ArttoSMart (System Modelled) interface followed by ongoing research work and Conclusions.

Data Modelling

We have not captured the existing choreography since we needed to generate something novel but within the classical framework. The following is the proposed model to represent a BN dance step. Various combinations of body limbs makeup a dance pose from amongst the legitimate poses. To model a BN pose, we need to represent the exact position of the 6 major limbs identified earlier. A Dance Position (DP) vector can be a combination of different attributes which were identified to model different body parts. The hands have 8 attributes while legs have 5 each. Two vector attributes were used to model head and waist attributes. Thus the DP vector is a 30 attribute vector corresponding to a dance position. Eyes depict emotions and are very expressive in a dancer. Eyes have no representation in the DP vector because we restrict the representation to *Nritta* that excludes abhinaya, excludes eye expressions. Similarly, the Neck can be modelled when depicting Head movements since they are fused together (although NS and AD clearly show 9 Neck movements depicting various movements). While modelling Hand movements, all the various sub-parts of a hand were considered: elbow, wrist, palm and the *hasta-mudras* (specialised gestures done by fingers). An example of a (*hasta*) *mudra* is shown in Fig. 2. Leg movements were modelled with the help of waist/ hip movement, knee and ankle positions. Thus the body parts are modelled using their sub-parts, with the relationships between them represented by X, Y and Z axes. The dance step of Fig. 1 is represented using the 30 DP vector shown below the figure, wherein the components are interpreted as follows:

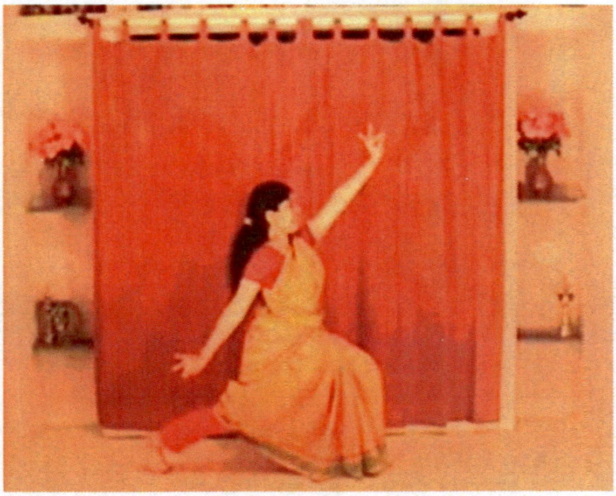

Fig. 1 An *Adavu* pose to depict a DP vector [0, 0, 3, 4, 2, 1, 0,1, 1, 0, 3, 0, 2, 0, 0, 1, 1, 0, 1, 0, 1, 0, 2, 0, 0, 0, 1, 1, 0, 0]

Limb	Attributes	Values
Head	[a1, a2]	[0, 0]
Right Hand	[a3, a4, a5, a6, a7, a8, a9, a10]	[3, 4, 2, 1, 0,1, 1, 0]
Left Hand	[a11, a12, a13, a14, a15, a16, a17, a18]	[3, 0, 2, 0, 0, 1, 1, 0]
Waist	[a19, a20]	[1, 0]
Right Leg	[a21, a22, a23, a24, a25]	[1, 0, 2, 0, 0]
Left	[a26, a27, a28, a29, a30]	[0, 1, 1, 0, 0]

A Hand movement from right to left represents X-axis. Thus all movements of the right hand on the right side are assigned positive values, and movements on left, negative. Similarly for left hand too. Y-axis is represented through upward (positive) and downwards (negative) movement. Front (positive) and back (negative) movements of the hand are represented through the Z-axis. Elbow movements can be straight to a bent elbow and thus values are assigned accordingly. Palm movements 1, −1 and 0 represent palm facing up, down and normal.

A shoulder pulled-in is represented by +1 and pushed-out by −1, whereas the normal shoulder position is 0. Leg positions in dance involves movement from the hip, position of the knee (half bent as in BN, full sitting or standing straight) and ankle positions. Similar to hand movements, here also right-side movements are positive and right leg towards left will be negative. A leg raised up or kept down is tracked through Y-axis and front and back movements through Z-axis. Refer [7] for more details.

The direction of head from downwards to upwards decides the values of head attributes, whereas for waist position we model a twist and a bend separately using two attributes.

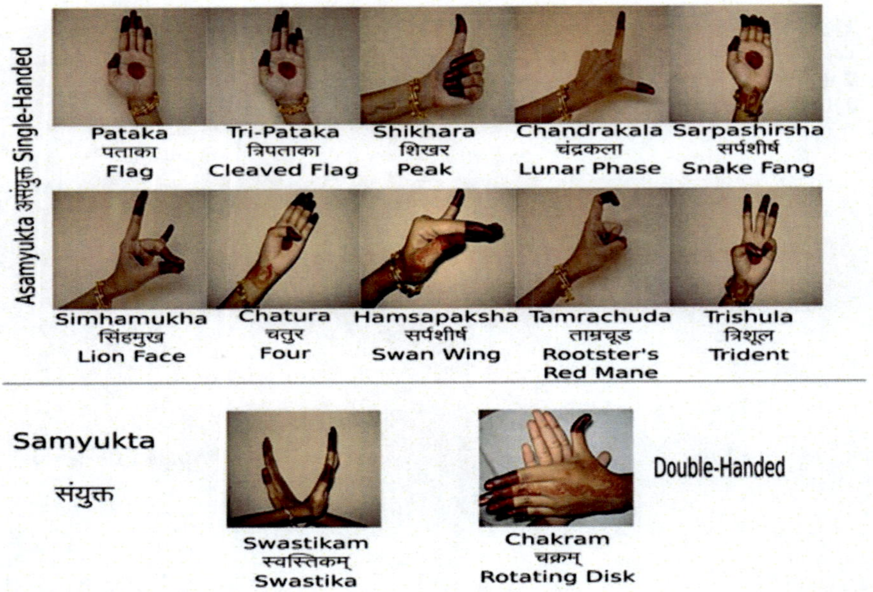

Fig. 2 *Hastamudras* or hand gestutes

Generation of New Dance Poses

Our objective was to generate unexplored dance steps without violating the BN framework. We have tried our level best to not deviate much from the original traditional style. Dance choreography needs a choice of combinations of thematic movements and motifs arranged in sequential and cyclic permutations that exhibit smooth continuity punctuated by beats and phase shifts. This obviously adds complexity to the grammar-driven complexity of verse and its continuous rhythmic musical rendering. We have shown earlier (see [7]): a simple set of elementary body movements (from the ancient texts of NS and AD) has the following complexity: 9 Head movements; 4 Neck movements; 8 Eye movements; 39 Leg movements; 28 single hand and 23 double hand movements. For choreography of each beat in each of these, the permutations and combinations will result in 8.6×10^{18} movement possibilities! The huge variety resulting out of all these possibilities renders computational modelling infeasible without an intelligent choice mechanism.

We have explored innovative choreography which remains in the BN domain but not practised before. Details are as follows: the huge combination possibilities for generating new poses, as we have just seen, had to be filtered out and hence we have used Genetic Algorithm for optimization. To express the aesthetic quality of dance steps that stand out from among zillions of possibilities, we have designed a fitness function. This fitness function determines a notion of distance of generated dance steps from the ideal steps, or elementary units of *Nritta* (*Adavus*). These dance steps have been represented using our DP vector model as described in Sect. 11.2.

Genetic Algorithm: Optimal Is Aesthetic

Genetic Algorithms (GAs) are local search heuristics that search the 'best' alternatives in the neighbourhood of some candidate solutions. These heuristics are best suited for applications to optimization of mathematically complex objectives. A GA starts from a random initial population; then carries out iterative improvement by using genetic operators such as mutation and crossover to create successive new generations. Variation is introduced by mutation and flattened out through crossover operation. To retrofit a GA to our problem of aesthetic design choice by computation, we had to mainly devise a suitable idea of fitness that expressed aesthetic preference. We achieved this through our notion of 'distance' from *Adavus*. For selection, fitness is determined through the intuitive distance measure. This selection process retains better dance steps from a set of enumerated dance steps of one generation to the next. We aimed to go beyond *Adavus* retaining their aesthetic value in dance choreography. Therefore, we built our distance notion on two parameters: one, limb variation count (LVC) allowing different steps maintaining the coordination and two, absolute vector distance (AVD). These parameters are presented below, earlier described in our publication [9].

LVC: It gives the count of body parts that are distinct in between two DP vectors. For example, if we have a change in hand and waist position between two DP vectors and the rest limb positions are exactly the same, then the limb variation count would be 2 and so on. More the value of LVC, further apart are the two dance vectors from each other.

AVD: gives the total variation distance between two dance vectors. A cumulative sum of absolute differences between the corresponding values of the thirty attributes of two DP vectors.

Thus, the distance 'd' is a function of AVD and LVC:

$$d = f(AVD, LVC) = (0.75AVD) + (0.25LVC) \tag{1}$$

Weight selection: We take AVD : LVC at 3:1 ratio to make the variation and novelty expressed in the new dance step proportional to the vector difference, while correlating limb variation with deviation from an ideal dance step. Thus the proposed fitness function behaves like a normal distribution curve. Fitness function value (FF) is given by

$$FF = f(d) = ND(d) \tag{2}$$

where ND gives normal distribution of the distance. ND ensures that higher fitness value shall be assigned to the dance vectors that are not too close or not too far from the ideal vectors.

Results The notion of distance of generated steps from the *Adavus* explicitly incorporated in the fitness function [9] took care of aesthetic propriety. However we faced one challenge: in-feasible steps were also generated in large numbers, and they were rendered fit by the function. To eliminate them from search, we maintained a database of dance steps that were in-feasible as a filter. Upon this foundation, we proceeded to perform automated classification as a natural next step.

Classification: High Dimension and Reduction

Expert opinions helped us in calculating the Mean Opinion Scores for evaluating the ratings for the dance poses. In [8] we reported: 'The system was trained up to 501 instances, using WEKA. Out of the 224 trained instances, we had 130 instances tagged as Excellent from Adavu data and 94 instances tagged as OK, Good or Bad based on an average obtained from the expert ranking. The classifier accuracy increased to 87.42% with these 500 instances which was far better than 66.96% for the 224 trained instances'. Finally we experimented with n-beat sequence generator for classification purposes and the details can be referred to at [10].

The experiment was limited in scale due to the obvious resource constraints, yet it provided substantial proof of the relevance of our generative combinatorial model. In order to scale it up, we needed to reduce dependence on human expert intervention or supervision, which is not only costly but scarce and uncertain or impeding—experts in a classical tradition are hardly amused by these ideas, which prima facie appear like the very antithesis of traditions and aesthetic sense.

Even with these clever tricks the search in GA was not very efficient, and the main challenge still remained the large size of the search space, and that in turn was because of the high dimension. A 30-attribute DP vector still allows at least 1 million possibilities, and considering that most attributes had more than two values, it would be 200 billion. But we must not take the high dimension to be realistic, considering that most of the limbs cannot move independently of others. Dimensionality reduction is a natural idea in such a situation. We address this next.

To make the automation of classifying these poses and for the best feature selection computationally feasible, we used Rough Set Exploration tools (RSES 2.2.2). This helped us in retaining high accuracy while selecting the best and minimal set of feature vectors [11]. We finally obtained 8 reducts from the 30 DP Vectors; however, this ultimately couldn't give very convincing results. The experimentation carried out with RST can be referred to at [10]. This approach, either using rough sets or any other dimensionality reduction technique, needs to be explored more.

Fractal Dimension for Aesthetics

Appreciation in any field of art like painting, drawing and sculpture, or even in the field of performing arts like dance, depends on *subjective* aesthetic sense. Aesthetic sense of an individual depends on several factors: it could be the various cognitive processing stages of the individual, which could in turn depend on different parameters like emotional quotient, perceptual analysis, familiarity, personal taste, symmetry, order and so on. To make it possible to automate the process of design and composition, we need some universal core embedded in subjective aesthetic sense. This core has to be captured in an objective and computable manner. One of our aesthetic measuring parameter experimentation was carried out with the Fractal Dimension (FD) for the system generated dance poses. The related literature review of why and how FD were chosen for this experimentation can be studied from our previous publication [12]. A brief discussion of the experimentation can be glanced here: the FD range for the already existing ideal *Nritta* steps, *Adavus*, were calculated. These are a benchmark since *Adavu* combinations are the only ones used for *Nritta* in a BN dance. Later, a sample size of 107 pictures generated by our system was chosen. These were rated by 9 experts from Pune and Goa on a Likert scale of 1 to 5 where 1 (Not Acceptable) was the least and 5 (Excellent) was the highest. We noticed that almost 93 pictures out of these 107 had a rating of 5 or 4 by at least one dance expert from among the 9 experts. But the ratings of all the images were not in the same manner by all experts due to obvious reasons. Hence these pictures were tabulated on the basis of being liked by more than 50% of experts. The final result was choosing a FD range for every expert as per their ratings and also proving that our hypotheses of *Adavus* being in the pleasing FD range. The detailed Statistical experimentation of proving our hypotheses along with the necessary images are published in [12].

Generating N-Beat Dance Poses for Choreography

We cannot directly apply the single beat choreography program for the N-Beat sequence. The results of directly using the same can be noticed in Fig. 3 where the hand *mudras* are constantly changing with no proper transition possibilities! It does not appear to be a seamless choreography and hence we need to use the following rule generation methods: We have attempted to simulate the *Adavus* which are considered the best for *Nritta* choreography.

Rule generation: An N-Beat sequence of several dance poses can be considered composable for choreography if it satisfies certain constraints. The filters or constraints identified to be applied to simulate *Adavus* (since they are the best for *Nritta*) fall under the following main heads:
Hand *Mudra* Filter: All the hand *Mudras* cannot have changes in them frequently. We need a smooth transition from one dance pose to another dance pose using these

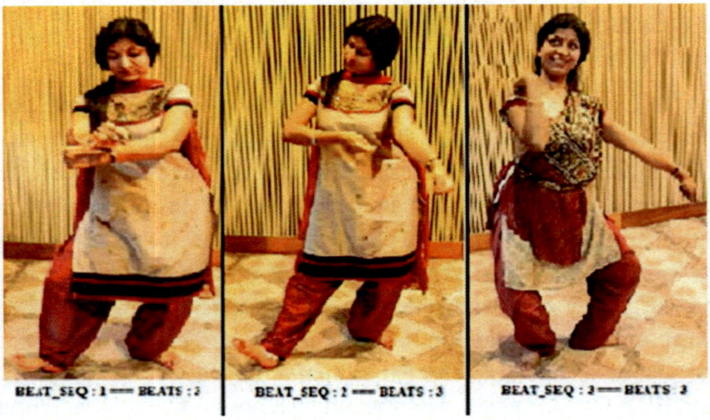

Fig. 3 3-beat sequence

hand *Mudras*. Plus, we need the whole sequence to come together aesthetically. Thus a hand *Mudra* cannot keep changing for every beat. Therefore, whenever the hand *mudra* is detected, the filter selects only those which are feasible possibilities from this *mudra*. Every hand *mudra* has only certain number of possibilities available. This detailed discussion can be found in [9]. We too restricted our system with the help of this filter accordingly.

Leg Filter: An *Adavu* does not look appealing if the leg position does not change for most of the beats. The dancer would appear stationary while only making hand movements. Hence this filter is another important feature.

Fitness Function Value Filter: This filter comprises of 3 parameters: Absolute Vector Difference (AVD), Fitness Function (FF) and Limb Variation Count (LVC). Refer equations (1) and (2). The single beat choreography results were also based on the same parameters as above. The same were also reproduced for continuity in multibeat results [13].

The 30 attribute DP vector was in a numeric form and deciphering the poses needed us to manually draw the same. Thus we requested a dancer to strike a pose of the DP vector and captured the pictures. To automate this manual work, we plotted human body notated stick figures as graphical plots using GNU Octave [14]. Although we got 80% accuracy, the system had its own fallacies due to intricate hand gestures and dimension loss and obviously a 3D system could help us better in this process.

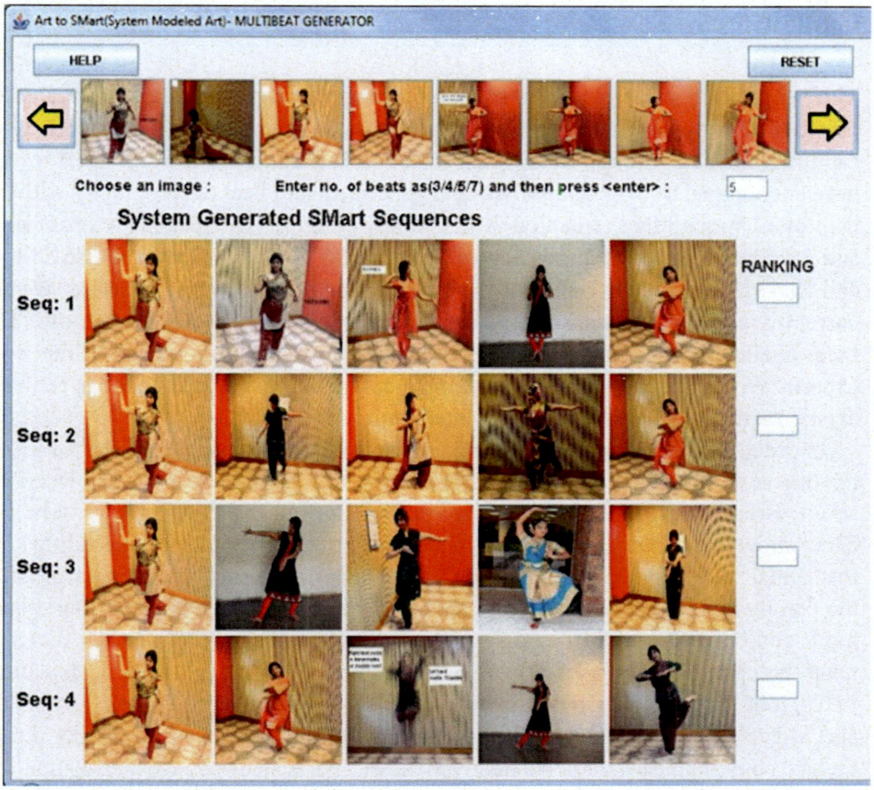

Fig. 4 5-beat choreography generated

The ArttoSMart Interface

We reported development of an architecture of a system to automate choreography as early as [11]. We have described above how we designed this to generate a multi-beat dance sequence not repeating the existing *Adavus*. Sequences generated by our combinatorial were reviewed to be novel and innovative, and commendable in choreography, by several dance experts. Given the starting pose, this multi-beat generator generates ten different N-Beat BN dance sequences. Every subsequent dance step in all these dance sequences is derived using all rules obtained in the rule generator module. The generated sequence was shown to experts in visual form. Snapshot for a 5 beat sequence in Fig. 4 helps explain the working of the multi-beat generator [11]. Our ongoing research on further computational experiments and analytical models is only trying to separate and compose the parts that can be automated and parts that require human expert knowledge into an organic cyclic process of design. Our two-pronged approach is what can be automated, try to model it as succinctly as possible, and what cannot be automated, try to make it easily accessible to human supervision.

Conclusions

This research is an unique attempt in suggesting the choreographer variations for pure dance movements. We have successfully modelled the human body instead of capturing the data through Kinect Camera or other methods. The Single beat results have encouraged us to develop filters and generate Multi beat choreography with the help of a Unique fitness function designed for the same. The results are promising, and we could reduce the dimension of 30 attributes to 8 with the help of RSES tool and also validate the aesthetics of the dance pose with FD as one measure. We are currently working on finalising the aesthetic parameters of a BN dance pose. The main challenge is getting curated datasets, due to lack of interest and time from experts. We are in the process of working a computational system that can evaluate existing media without much expert intervention.

In order to reduce this dependence on expert supervision, we have started investigating aesthetic universals. Our experiments with fractal dimensions are described above. However, fractal dimensions are suitable for images and singular poses. For dance movements, we had to search broader and deeper. We found the leading neuroscientist V. S. Ramachandran's work [15] most promising in this regard. After the completion of the doctoral work described in this paper, we started investigating if and how Ramachandran's nine principles of aesthetics ("*Navarasas*") can be made computable. These principles are Grouping, Peak shift, Contrast, Isolation, Perceptual Problem-Solving, Abhorrence of Coincidences, Orderliness, Symmetry and Metaphor. We are investigating if and how to build a basis for creating a full-fledged computable aesthetic model from these. This is an ongoing experimental and analytical study. This ongoing work is presented in another chapter in this volume.

Acknowledgements I would like to thank my adviser from Goa University, Dr. Jyoti D. Pawar for all her timely help and suggestions and Dr. Manish Joshi, Associate Professor from North Maharashtra University who helped me immensely with this UGC sponsored Major Research Project (39- 901/2010(SR)).

References

1. S. N. Pattanaik, A stylised model for animating Bharatanatyam, an Indian Classical Dance form. Comput. Art, Des. Animat. (Springer, 1989)
2. A. Mallik, S. Chaudhury, H. Ghosh, Nrityakosha: preserving the intangible heritage of Indian classical dance. J. Comput. Cult. Herit. **4**(3), 11:1–11:25 (2011). https://doi.org/10.1145/2069276.2069280
3. S. Jadhav, A. Aras, M. Joshi, J. Pawar, An automated stick figure generation for Bharatanatyam dance visualization, in *Proceedings of the 2014 International Conference on Interdisciplinary Advances in Applied Computing* (ACM DL, Amritapuri, Coimbatore, India, 2014). https://doi.org/10.1145/2660859.2660917
4. R. Majumdar, P. Dinesan, Framework for teaching Bharatanatyam through digital medium, in *2012 IEEE Fourth International Conference on technology for Education* (IEEE Computer Society, Hyderabad, 2012), pp. 241–242

5. B.S. Anami, V.A. Bhandage, Suitability study of certain features and classifiers for Bharatanatyam double-hand mudra images. Int. J. Arts Technol. **11**(4), 393–412 (2019)
6. T. Mallick, P.P. Das, A.K. Majumdar, Posture and sequence recognition for Bharatanatyam dance performances using machine learning approach. arXiv preprint arXiv:1909.11023 (2019)
7. S. Jadhav, M. Joshi, J. Pawar, Modelling Bharatanatyam dance steps: art to smart, in *Proceedings of the CUBE International IT conference & Exhibition Proceedings* (ACM DL, PUNE, 2012)
8. S. Jadhav, M. Joshi, J. Pawar, Art to SMart: an automated bharataNatyam dance choreography. Taylor & Francis **29**(2), 148–163 (2015). https://doi.org/10.1080/08839514.2015.993557
9. S. Jadhav, M. Joshi, J. Pawar (2012) Art to SMart: an evolutionary computational model for Bharatanatyam choreography, in *IEEE Xplore* (2012), pp. 384–389. 978-1-4673-5115-7
10. S. Jadhav, J. Pawar, Bharatanatyam dance classification with rough set tools, in *ICT Based Innovations*. ed. by A.K. Saini, A.K. Nayak, R.K. Vyas (Springer Singapore, Singapore, 2018), pp. 75–81
11. S. Jadhav, M. Joshi, J. Pawar, Towards automation and classification of Bharatanatyam dance sequences. Techno-Math. Res. Found. **11**(2), 93–104 (2014). http://www.tmrfindia.org/ijcsa/v11i27.pdf
12. S. Jadhav, J. D. Pawar, Aesthetics of Bharatanatyam poses evaluated through fractal analysis, in *Proceedings of the First International Conference on Computational Intelligence and Informatics*, ed. S.C. Satapathy, V.K. Prasad, B.P. Rani, S.K. Udgata, K.S. Raju (Springer Singapore, Singapore, 2017), pp. 401–409
13. S. Jadhav, M. Joshi, J. Pawar, Art to smart: automation for Bharatanatyam choreography, in *19th International Conference on Management of Data* (CSI, India, Ahmedabad, Gujarat, India, 2013), pp. 131–134
14. S. Jadhav, A. Aras, M. Joshi, J. Pawar, An automated stick figure generation for Bharatanatyam dance visualization, in *Proceedings of the 2014 International Conference on Interdisciplinary Advances in Applied Computing* (ACM DL, Amritapuri, Coimbatore, India, 2014). https://doi.org/10.1145/2660859.2660917
15. V.S. Ramachandran, *The Tell-tale Brain: A Neuroscientist's Quest for What Makes us Human* (WW Norton & Company, 2012)

Interfacing the Traditional with the Modern

Automatic Mapping of BharatNAtyam Margam to Sri Chakra Dance

Ambuja Salgaonkar, Padmaja Venkatesh Suresh, and P. M. Sindhu

The first half of the paper consists of an overview of the research in computer-assisted choreography, an introduction to the Sri Chakra design and a description of the BN protocol with necessary details of the dance form, while computational experiments and their outcomes are discussed in the second half. Possibilities of employing automation for enriching the activities in the dance domain have been elucidated and a few pointers for further research are provided.

Preamble: Automation in dance, a 50 year old domain of research, has facilitated four aspects [1]: (i) *dance capturing,* i.e., recording information about a dance performance for analysis and reproduction, (ii) *dance understanding,* i.e., uncovering syntax and semantics of gestures in order to interpret a performance, (iii) *dance making,* i.e., articulation of dance steps and choreography, and (iv) *dance applications,* i.e., producing a dance for a specific purpose like entertainment [2], tutoring [3–8] or therapy [9].

The paper suggests an approach for computing the transitions in the Sri Chakra (SC) dance, a theme in which the Hindu goddess Shakti or god Shiva is worshiped in the form of a diagrammatic representation that is also interpreted as a symbol of the universe. In this dance, the dancer is a devotee, yearning for union with the deity, the

A. Salgaonkar (✉)
Department of Computer Science, University of Mumbai, Mumbai 400098, India
e-mail: ambujas@udcs.mu.ac.in

P. V. Suresh
Aatmalaya, Bangalore, India
e-mail: padmajasuresh@hotmail.com

P. M. Sindhu
Nagindas Khandwala College and Department of Computer Science, University of Mumbai, Mumbai, India
e-mail: sindhu.pm.satheesh@gmail.com

A. Salgaonkar and M. Velankar (eds.), *Computer Assisted Music and Dramatics*, Advances in Intelligent Systems and Computing 1444,
https://doi.org/10.1007/978-981-99-0887-5_12

cosmic principle. Therefore, it is customary to refer to the SC dancer as a *devotee*. Throughout the paper, the two words are used synonymously.

SC is an assembly of nine concentric designs called *Mandalas*, i.e., circles, or *Avaranas*, i.e., covers. The dance demonstrates *Margam,* a 9-step protocol for elaborating a performance in *Bharatanatyam* (BN), a variety of south Indian classical dance. Details are termed *NavAvarana-kritis*, literally, to dance in the nine Avaranas. The philosophical basis to the BN MArgam (BNM) of the Thanjavur quartet has been inspired by eighteenth century poet musician Sri Muthuswamy Dikshitar [10]. It follows a traditional Hindu temple architecture that in turn replicates the SC.

So far it has been understood that the traversal of SC is a journey of a dancer at her mental level. This is because having the structure physically painted and traversed on the dance floor has limitations: (i) dancing on the actual temple floor is restricted by the installations inside the temple and (ii) drawing the SC is a complicated process [11–14]. Also, drawing the SC within the dimensions of a given stage or in which a dancer is comfortable is a challenge and hence is a research problem.

A variety of algorithms have been devised and demonstrated for drawing and traversing SC by employing concepts from geometry and graph theory [15]. Tuning them to the actual requirement of a BN dancer and providing automated drawings of the SC consistent with the given dimensions is an objective of this paper. Measurements have a special importance in SC drawing. Dancing according to the printed design could enhance performance accuracy, and witnessing a dancer traversing such a complicated design would be a treat for the spectators. Interpretation of the BNM with respect to the NavAvarana-kritis has been considered and documented here for the first time. Putting together the automated drawing of SC and the interpretation of BNM, a computer-assisted choreography for the SC dance has been proposed as a new research problem.

In summary, an interpretation of traditional BNM as Sri Chakra dance along with a computational model for the spatial specification of the dancer's movements on the floor is a contribution of this work. To the best of our knowledge, this is the first time in the history of Indian classical dance, the Sri Chakra theme has been considered for computer-assisted choreography.

We start by giving a brief account of research in dance making. Next, we give a brief introduction about the SC diagram and about the *Chakras*, esoteric energy centers along the spine that are invoked in the SC dance. Thereafter the authors' interpretation of the BNM as an SC dance follows.

In the later part of the paper, an algorithm for drawing the SC and two algorithms for its traversal are provided, followed by a discussion on their application in practice and probable value additions. An exploration of a computer-assisted choreographic system is provided in the concluding part.

Glimpses of Research in Dance Making

Dance making involves designing a step, a sequence of steps, or transitions between the steps and step-sequences, that would appeal to spectators. Two approaches are seen in the literature. One, a system reads a *notation* of a dance piece, i.e., systematically documented information about the gestures-in-series and corresponding music. A computer-assisted interface for it is provided in [16]. Two, a system senses human performance and creates visuals, either on a monitor [17] or a dancing humanoid robot [18] or both [19]. Alternatively, a system analyzes a performance corpus and, by employing machine learning techniques like genetic algorithms, synthesizes a novel piece of dance that is acceptable to the tradition [20, 21]. The greedy approach to the problem of dance choreography has been compared to genetic algorithms [22]. While domain ontology has been employed with deep learning neural networks or support vector machines for segmentation of a dance performance, the hidden Markov model has been used in dance synthesis [23, 24]. A transition has been visualized as an edge joining two vertices of type *posture*, i.e., an instantaneous body state [25]. Hagendoorn [26] is a rule-based system and [27] employs genetic algorithms for estimating possible graceful interactions among a group of dancers on a given stage and hence suggests movements for individuals, while [28] employs interpolation for defining inter-posture transitions. Six of these references are on Indian classical dance. More research in this domain is called for.

Sri Chakra Diagram and SC Dance

The core of SC is primarily a matrix of nine interlocking triangles. Five triangles (corners painted in blue, red, black, green and yellow) with their apex facing downward represent Shakti. Four triangles (corners painted in blue, red, black and green) with apex facing upward represent Shiva. The assembly is placed in a bloomed lotus having two rounds of petals (painted in red and yellow), which is placed on a sacred platform painted in green (Fig. 1). Note that the color scheme is not universal.

The intersection of the nine triangles in SC creates forty-three, or by considering the center as a triangle, forty-four triangles. The image could be perceived as an assembly of nine mandalas. Figure 2 gives a mandal-wise labeling of these triangles from 0 to 42; the labels in one mandala are in one color.

Traditionally it is believed that these Avaranas relate the human constitution in its physical, mental and vital levels to the nine-fold cosmic energy. The number 9 also signifies the *NavaRasa*, the nine aesthetic moods, namely, *SringAra* (love), *Veera* (Valour), *HAsya* (humor), *Karuna* (compassion), *Bhibatsa* (disgust), *Adbhuta* (wonder), *Bhayanakar* (fear), *Raudra* (fury) and *Shanta* (peace).

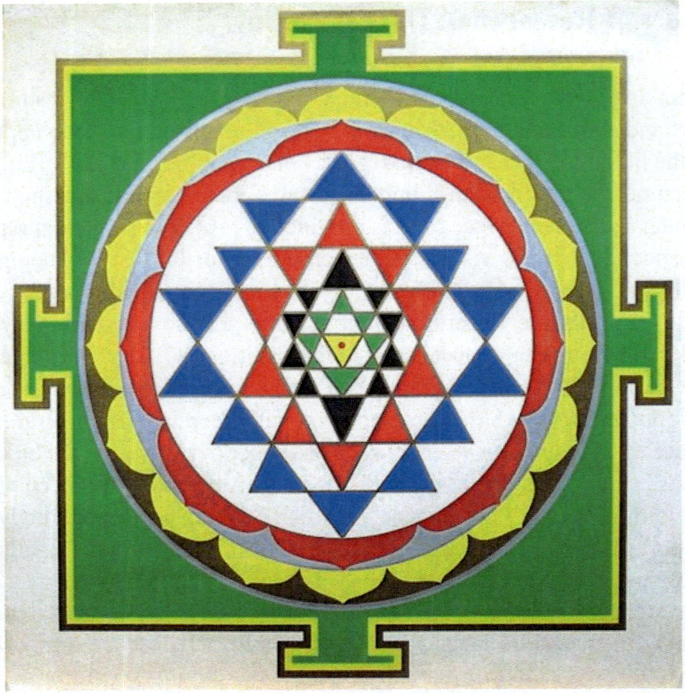

Fig. 1 Sri Chakra, *Source* [29]

Sri Chakra has been worshiped by employing absorption or dissolution in Bhāvanopanishad, affiliated to Atharva Veda, also termed *Laya yoga.* A devotee begins from the outermost circle in the design and systematically proceeds inwards. She returns to the starting point after reaching the center, the *Bindu* (a metaphysical point that symbolizes concentration of luminous energy with the natural instinct for creation). The worshipers associate themselves with *kriya* (a yogic technique), *mantra* (sacred sound), *mudrA* (a mystic symbol) and *KaranAs* (specific transitions from Natyashastra, the traditional text on dance and drama) [30] that invoke the spirituality within the dance spectators.

The geometric patterns serve as a scaffolding to the dance. The mystic images in the yantra determine the motifs and movements. The triangles, squares and circles in it start falling in place, like individual instruments in a musical ensemble. A triangle is the most stable architecture. The triangles with the definite positions of the bent knees and head, give stances to a BN dancer, namely, *Purna and Ardha Mandali* (full and half sitting postures). The dance sequences are seen to be resembling the Yantra *Mandalas.*

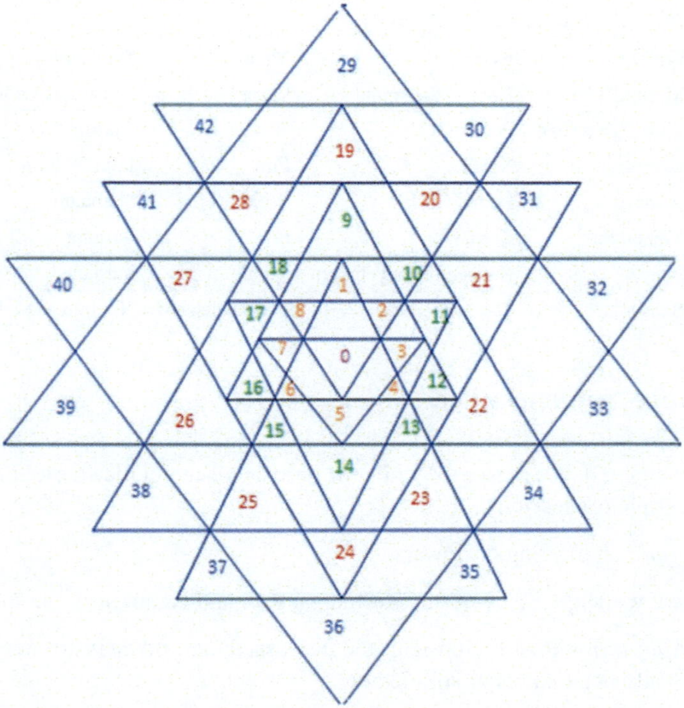

Fig. 2 Mandala-wise labeling of SC triangles; *Source* [15]

Chakra-Philosophy

There are said to be seven chakras in a body and it is believed that a chakra could be activated by meditating upon it. Chakra [31] gives an account of the acceptance of this philosophy as well as skeptical responses. *Prithvi* (Earth), *Aap* (Water), *Tej* (Fire), *Vaayu* (Wind) and *AakAsh* (Ether), the basic elements that make up any living organism, are called the *Pancha-Mahabhootas*. Mahabhoota-Chakra is a one-to-many mapping [32]. During the Sri Chakra dance the Chakras are implicitly meditated to invoke the energy around the corresponding region of the body. Details have been listed in Table 1.

With every progression in the conventional dance or Margam, the previous Chakras are believed to remain open and hence it is not a watertight compartment or division arrangement.

Table 1 Chakra and associated Mahabhoot

Sr No	Chakra	Region	Mahabhoot	Place in dance
1	MoolADhAr	Root of the spine	Earth	Invocatory dances
2	SwADhiShThan	Pelvis	Water	Alarippu
3	Manipura	Navel	Fire	JathiSwaram
4	AnAhat	Heart	Wind	Shabdam
5	Vishuddha	Throat	Ether	Varnam
6	AjnyA	Between the eyebrows	Mind	Varnam
7	SahasrAr	Top of the head	Consciousness	Not indicated, but implied

SC Dance Nomenclature

*MallAri***:** a pure dance presented while the deity is placed in the chariot or palanquin for the temple procession.

PushpAnjali: an offering of flowers.

Kavutvam: worshipping Ganesha, Kartikeya, Kali and Nataraja.

MelaprApti: a rhythmic recital with the percussive and other instruments with the leading Nattuvanar's bols or sollukkattu.

Alarippu: a bud blooming into a full lotus.

Alapadma: a lotus blooming.

PatAka: a hand gesture denoting a flag, water waves, wind, expressions like me, you, etc.

RAga: a melody framework of the South Indian classical music.

TAl: a rhythmic cycle, conforming time with hand claps.

Swaras: notes of a raga.

Nritta: a non-representational pure dance.

Jathi: a sequence in Nritta.

Adavus: standard basic steps in BN.

Korvais: a series of Adavu combinations.

Abhinaya: harmoniously enacting the semantics of the poetry leading to *Rasa,* aesthetic delight.

PadaVarnam: the quintessence of BN in which feet and face enact simultaneously.

PoorvAngam: the first part.

UttarAngam: the second or latter part.

Pallavi: the opening of a lyrical composition with variations in phrase and tempo.

Anupallavi: follows Pallavi in order to add further meaning and beauty.

MuktayiSwaras: melodic notes with dance to which lyrics follow in consonance.

SAhityams: lyrics.

Charanam: one-line lyric.

CharanamSwarAs: three to four couplets consisting of Swaras taking lyrics to a climax.

Teermanams: literal meaning is conclusion; advanced patterns presented in the latter part of Nritta.

NAyikABhAva: the dancer-devotee imagines herself as the heroine of the deity.

Angika Abhinaya: expressive body movements.

VAchika Abhinaya: acting through speech.

SAtvika *Abhinaya*: presenting emotions through facial expressions.

Mei Adavu: a collection of steps where the body parts are gracefully rendered in gradual rhythmic renditions, e.g., the eyes, shoulders and torso are moved with the feet.

BNM and SC

Each of the 9 steps of the BNM has been described from the dance point of view. The references to a few philosophical terms may be outside the domain of the modern sciences. However, the authors believe that knowing them is essential for the tradition of BN.

1. **Invocatory dances**: They begin with the dancer's entry through the 9th or outermost Avarana, Bhoopuram (BP), the gates marked in the square boundary. The entry signifies the beginning of SC worship from the gross earth element by invoking the Mooladhara chakra. NAdaswaram and Tavil accompany this dance. Tadi tom nam, Dhim Dhim and other syllables are lyrically uttered. These mnemonic syllables, reflecting the pure light energy, are a specialty of these dances. During PushpAnjali, a praise is rendered to Lord Ganesha for the removal of obstacles. Shiva or any deity of the Shaivite pantheon is worshiped with praise as a Kavutvam is carried out. Thereafter the MelaprApti begins. The nine gods (Brahma, Agni, Indra, Varuna, VAyu, Kubera, Nirutti, Yama and IshAna), representing nine directions, are praised through Navasandhi Kavutvams.
2. **Alarippu**: This step marks the dancer's entry into the 8th Avarana that touches the sixteen petal lotus, beyond the concentric circles called Veethi. The dance gradually starts with charming movements of eyes, neck, hands and feet. Alapadma

(basic hand gestures) follow PatAka that symbolizes a temple gate- Gopuram. This step displays geometric movements in accordance with the drum syllables. The rhythm consists of combinations of 3 and 4 beats. The focus is on the SwadhisthAn chakra.

3. **Jathiswaram**: The dancer enters the 7th Avarana that is outside the eight-petal lotus. The music contains passages of Swaras. The focus is on the Manipura Chakra. This is an item of Nritta followed by Korvais. The number four that stands for the squares, *Vedas* (the holy texts) or *VAnee* (forms of speech), is established. Four expands in eight or sixteen (petals). The number three symbolized by the circles represents the three powers, creation, sustenance and dissolution, or the three states of mind, JAgrut (awake), Swapna (dream) and Sushupti (deep sleep); or the three Gunas, characteristic of nature, Sattva, Rajas and Tamas. Four combined with three leads to seven that expands into fourteen which is the number of triangles in the next Avarana.
4. **Shabdam**: The literal meaning of Shabdam is voice. Here it refers to the lyric praising the deity or a patron. The song is interspersed with small Jathis. Abhinaya is expected to begin here. The dancer enters the 6th Avarana at the boundary of the assembly of 14 cognate triangles. The number 14 symbolizes Panini's *Maheshwar sutras* (the first 14 rules of Sanskrit grammar given by Panini). The meditation is upon the Anahata Chakra. The number seven in fourteen symbolizes the *Dhatus* (life sheaths) and *MatrikAs* (divine mothers).
5. **Varnam**: Varnam means color, as also the single seed syllable that vibrates continually in the heart of the devotee. Here, it refers to the color at the grand entry of the dancer into the subtler levels from 5th to the 1st, i.e., Bindu. The dance consists of Nritta and Abhinaya. It begins with four Jathis and is called PadaVarnam.

   Poorvangam consists of Pallavi, Anupallavi, MuktayiSwara along with SAhityam. Teermanam is executed at the end of Pallavi and Anupallavi. The devotee touches the Bahirdasha (outer ten) and soon after that the Antardasha (inner ten) triangles, i.e., into the 5th Avarana and subsequently into the 4th one. The number 10 represents vital air. The meditation is on Vishuddha Chakra.

   Uttarangam consists of Charanam followed by CharanamSwara. This marks the entry of the dancer in the 3rd Avarana, i.e., the outer boundary of the assembly of 8 cognate triangles. Seven of them represent the alphabet and the eighth represents manas—buddhi-ahankar-atirahasya yoginis (mind, intellect, ego and the goddesses of secrecy). The song increases in tempo, the following two passages go with speedy movements and stronger expressions that indicates the intensified yearning of the devotee to get united with the deity. She enters into the 2nd Avarana, i.e., the path directed by the Trikon Chakra. The five Chandrakalas, (crescents of the moon) are perceived by the dancer on each of its sides. Therefore, fifteen facets of the moon, collectively called Chandrakalas, are worshiped. The meditation is on Aajnya Chakra. Throughout, the dancer depicts NAYikAbhAva first by acting upon PadArtha (the words of the lyric) and next, VakyArtha (inducting ideas and situations that are not inherent in the text, but are woven around the subject matter).

   Nritta Korvai is performed for the MuktayiSwarA and CharanaSwarA.

Finally, the Varnam ends inside the Trikona chakra, close to the Bindu, i.e., the first Avarana within which the point of union with the deity, the place to salute the SahasrAr Chakra, resides. However, perhaps to indicate the abstraction or ideation within the intellect, beyond the ken of material limits, the union with the deity is often left alone by composers, so as to kindle the imagination of spectators. This is the mainstay of aesthetic bliss or Rasa, the intention of any artistic enterprise which is evoked by the master-artiste and not explicitly depicted. The Abhinaya in Varnam has all three aspects: Angika, VAchika and SAtvika.

6. **Padam**: It is the dance with deep emotional enactment, indicating the dancer's urge to hang on close to the Bindu. It is either the dance of *NatarAja* (Shiva) or the Nayika depicting her various moods while in a state of deep love and surrender. Padams give exclusive importance to Abhinaya. The music is slow and consists of Pallavi, Anupallavi and Charanam.
7. **JAvali**: JAvalis are lighter forms of Padams, in music as well as in theme. Their lyrics are colloquial and direct. Here the Lord is treated more like Nayika's companion. Hence, the Abhinaya is lighter, bolder expressions are used. The tempo is faster, the depictions are more realistic as compared to the previous ornate Padams. These love songs signify a descent from subtler to grosser levels.
8. **Keertnam**: It is devotional music addressed to various deities. The Abhinaya is Bhakti-based. The ancillary deities are remembered. The elaboration of the dance is based on episodes and stories from Hindu mythology. The metaphysical correspondence is with the movement back towards the petal structures.
9. **Tillana**: This item is dominated by Nritta. Beautiful rhythmic patterns are woven together, highlighting aesthetic dance movements. The music structure of Tillana has Pallavi, Anupallavi and Charanam. The dance has exclusive Mei adavus that include graceful movements, sculpture-like poses and rhythms with four, three, seven, five and nine beats. Tillana has RangAkramana movements around the stage, depicting the Pradakshina (circumambulation) of the SC along the flowering circles. Towards the end, there is a short SAhityam addressed to a deity for which the Abhinaya is performed. The SAhityam ends with the stamp or mudra of the music composer.

Shlokam/Mangalam: The recital generally ends with a Mangalam, a short devotional prayer in praise of the Supreme, seeking His blessings.

The SC dance takes the spectators on a journey from the gross world to the subtlest and back. The performer, music, lyrics, accompanists and spectators together produce the Rasabrahman, the climax of the aesthetic mood while the dancer gets back to the BP.

Drawing the SC

Recall that the SC comprises of three concentric parts (Fig. 1), namely, (i) the Central Assembly of 9 interlacing Triangles (CAT) that is embedded in (ii) two rounds of adornment of 8 and 16 petals (call, E and H) respectively, and (iii) the outermost Avarana, the Bhupuram (BP) that defines the boundaries of the design, a base of the structure (Fig. 1). Veethi (Vt) surrounding H could have been a circular track for the worshipers to orbit the core of the yantras. Therefore, it might not have been considered as a constituent of the 3-part construction and it is not considered explicitly in the dance. However, Vt could be useful as a heuristic for judging the distance between two petal-peaks of H while defining an SC dance choreography.

A MalayAlam mnemonic to draw SC has been produced in [15]. We generated a refined version by incorporating some optimizations and generalizations. The following algorithm is the result of this exercise.

Algorithm DrawSC

Step 1: Draw a circle C of radius 24 units.

Step 2: Let TB be a diameter of C that is parallel to the vertical axis; T and B are its top and bottom endpoints.

Step 3: Draw horizontal line segments CS1 to CS9 keeping their midpoints on TB, and their lengths and distances from point T as given in Table 2.

Step 4: Connect (1, M6); Connect (2, M9); Connect (3, B); Connect (4, M8); Connect (5, M7); Connect (6, M2); Connect (7, T); Connect (8, M1); Connect (9, M3).

Here Mi is the midpoint of CSi and Connect(i, j) joins the end points of the line segment CSi to j, thus forming a triangle with the base CSi and apex j.

Table 2 Lengths and locations of CS1 to CS9

Line Seg	Length	Distance From T
CS1	25.7	6
CS2	31.6	12
CS3	45.9	17
CS4	15.3	20
CS5	6	23
CS6	15.6	27
CS7	46.5	30
CS8	33.6	33.6
CS9	23.7	42

Step 5: E = DrawPetal (24, 8, 11); H = DrawPetal (35, 16, 9).

Here DrawPetal(r, n, h) draws a structure of n petals on the circle with radius r with respect to the center of the design, with petal height h.

Step 6: Vt = DrawVt (44, 1, 2).

Here DrawVt (r, i, n) draws a circle with radius r with respect to the center of the design, increases the radius by i, draws the next circle and repeats this process so that a total of n + 1 circles are drawn.

Step 7: BP = DrawBP (48, 13).

Here DrawBP (r, o) draws a square circumscribed by an imagined circle of radius r around the center of the design. In the middle of each side a segment of length o is deleted, thus visualizing four gates in the structure. The border is demarcated with three lines one unit apart.

For persons interested only in drawing SC and not in further mathematical modling the necessary arguments values of DrawSc have been provided in Table 2.

DrawSC algorithm is an instance of the generalized algorithms for drawing the Chakra-type designs [15]. The parameters required by each function to draw the complete assembly of a Chakra are automatically computed on the fly once the dimensions of the frame in which the design is to be inscribed are specified.

The following Fig. 3 is the output of DrawSC the CAT rendered in black. The next problem to be solved is SC traversal, for a choreographer to help mapping of the BNM on this geometry. The blackened triangles bring to the reader's notice that the difficulty of SC traversal is essentially due to the complexity of traversing this blackened structure. Traversal of the 2 petal structures and of the path inside the BP and around the petal structures are comparatively easier tasks.

Traversal of SC for BNM

First, we give details of our graph abstraction of SC applicable to similar Chakra-type concentric designs. Next, we analytically prove the existence of such a mapping and provide the exact traversal path.

Essential terminology is provided for novice readers of graph theory.

In mathematics, a *graph* G is defined as G = (V, E), where V is a non-empty set of vertices and E is the set of edges of the form e = (vi, vj), denoting that the edge e connects the vertices vi and vj and therefore vi and vj are *adjacent* vertices or *neighbrs.* A *walk* in a graph is a sequence of edges to be followed for traversing from source to destination. An *Eulerian Circuit* (EC) is a walk that includes all the edges of the graph exactly once and ends at the starting vertex. A graph containing an EC is *Eulerian graph.* The *degree* of a vertex is the number of edges incident on it. It is a theorem that a graph is Eulerian, if all its vertices are of even degree [33].

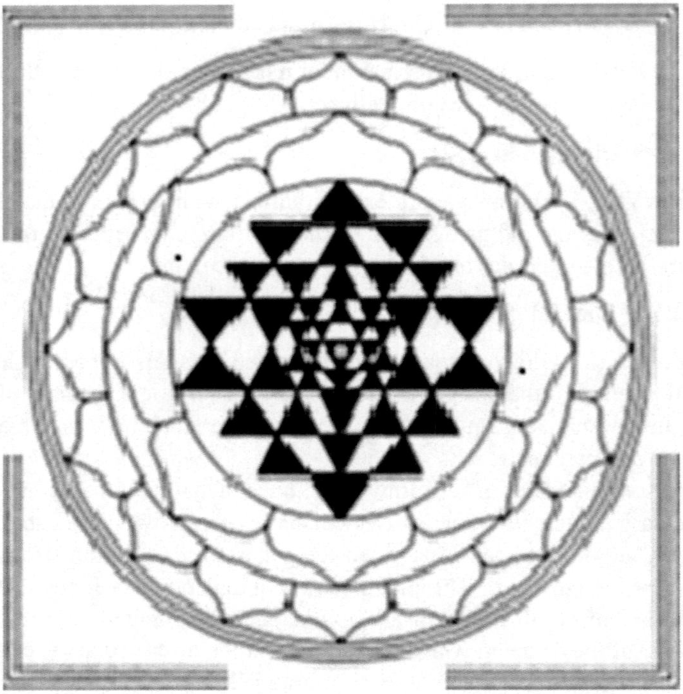

Fig. 3 Sri Chakra

A *planar graph* is a graph that can be drawn on a 2D surface without crossing edges.

A walk in which no vertex appears more than once is called a *path.* If there exists a path from v_i to v_j for every pair of vertices (v_i, v_j) in G, then G is called a *connected graph.* A path that starts and ends at the same vertex is called a *cycle.* A graph having no cycle is *acyclic.*

Chakra graph (CG): A Chakra could be perceived as an arrangement of vertices and edges—a graph, in other words. This perception will require some relaxations, adjustments and constraints, e.g., length measurements, orientation and sequence of drawing the image. A point of intersection or touching of the 9 interlocking triangles forming CAT and E and H (fundamental elements of the design) has been identified as a vertex in CG, ensuring that CG is a planar graph. SC is a connected graph whose subgraphs are chakra graphs *of CAT (CATG),* Vt and BP.

However, in BNM, the Vt being insignificant in the current context we need not consider the CG of BP in the formation of CG of SC. We call this abstraction of SC as SCG.

Figure 4 is a CATG in which the vertices are labeled layer-wise, starting with the outermost layer and going inward till the Trikona Chakra. This labeling is in line

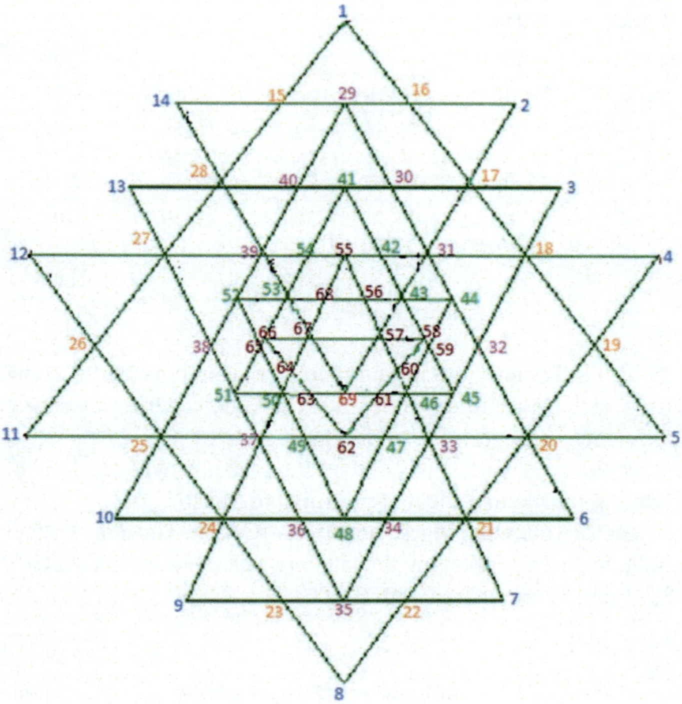

Fig. 4 *Chakra Graph of CAT in SC; Source:* [15]

with the direction of progress in the SC dance. Note that the base is considered as a separate layer on which a triangular or petal structure is installed.

Theorem: There Exists a Mapping of BNM with SCG

Proof: From the discussions in the section on traversal of SC for BNM it is clear that in BNM the path taken by a dancer in the ascent phase starts at a point in BP and reaches the Bindu by passing sequentially through the next 8 Avaranas from 8th to 1st. The path during the descent starts at the Bindu and reaches the starting point in BP by crossing the 8 Avaranas in reverse order from 2nd to 9th. This sequence of dance movements indicates the existence of an EC in the SCG. We compare this observation with the problem of finding an EC in a given graph, which is equivalent to proving that SCG is Eulerian.

We give a 3-step proof of this.

(i) CATG, consisting of 155 edges joining 69 vertices, forms 60% of the SCG structure. By visual inspection of Fig. 4, summarized in Table 3, we confirm that all the vertices in CATG are of even degree.

Table 3 Vertex degree and labels

Vertex degree	Vertex label	Total vertices
2	1 to 14	14
4	15, 16, 19, 22, 23, 26, 29, 30, 32, 34, 35, 36, 38, 40, 41, 42, 44, 45, 47, 48, 49, 51, 52, 54, 55, 56, 58, 59, 61, 62, 63, 64, 65, 66, 68, 69	36
6	17, 18, 20, 21, 24, 25, 27, 28, 31, 33, 37, 39, 43, 46, 50, 53, 57, 60, 67	19
Total		69

Thus, CATG is Eulerian and therefore, it is possible to construct a walk starting with an arbitrary vertex V in CATG, covering each edge of CATG exactly once. and returning to the starting vertex V.

V could be a vertex with label 1 or label 8 without loss of generality, depending on whether the deity is Shiva or Shakti.

(ii) Consider the following Fig. 5, i.e., SCG—CATG. The new vertices are labeled from 70 to 117 by following the labeling scheme used in CATG. Vertices 1, 4, 5, 8, 11 and 12 are shared with CATG

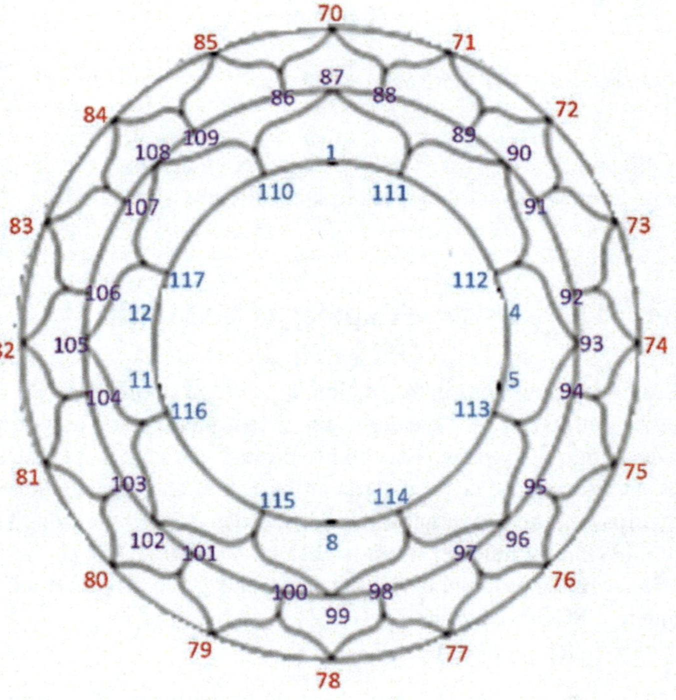

Fig. 5 SCG-CATG representation

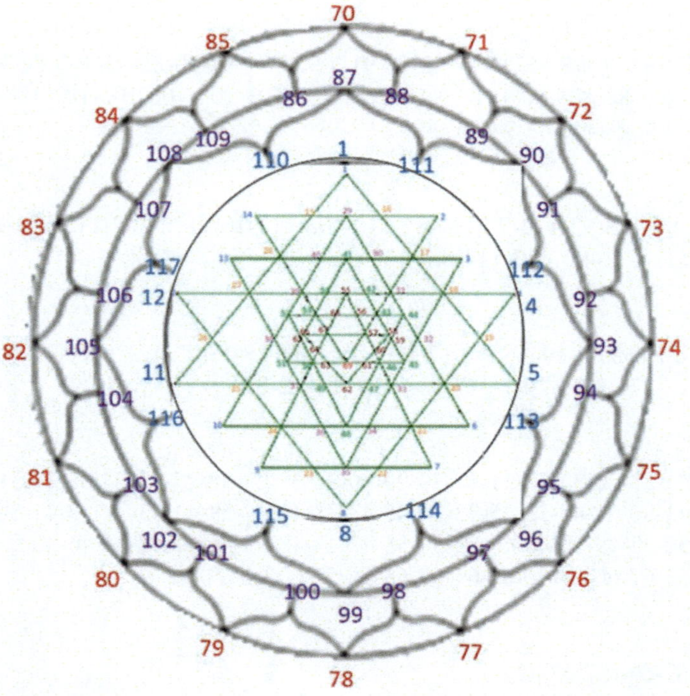

Fig. 6 EC in SCG

The vertices that are exclusive to this graph are of degree 4, while vertices common to this graph and CATG have degree 2 here. This means that all vertices in the graph SCG—CATG are of even degree. It follows that the graph SCG-CATG is Eulerian.

(iii) Select a vertex V that is common to both G1 and G2. Starting with V, traverse an EC in G1. Next, from V, start traversing an EC in G2. In completing the traversal, V is reached again, and all the edges of graph G have been traversed exactly once. Therefore, the total walk is an EC in G, i.e., G is a Eulerian graph.

Both CATG and SCG—CATG are Eulerian graphs. Hence SCG is a Eulerian graph.

An EC in SCG is given below that starts with vertex labeled 78, enters the CATG at point 8 and returns to 78. Observe that the lines drawn In Fig. 6 are either thin or thick. The thick lines convey the gross aspects of the dance and its philosophy while the subtle aspects are indicated by thin lines.

S1:

78-79-80-81-82-83-84-85-70-71-72-73-74-75-76-77-**78.**

S2:

78-98-77-97-76-95-75-94-74-92-73-91-72-89-71-88-70-86-85-109-84-107-83-106-82-104-81-103-80-101-79-**100**-101-102-103-104-105-106-107-108-109-86-87-88-89-90-91-92-93-94-95-96-97-98-**99.**

S3:

99-114-96-113-93-112-90-111-87-110-108-117-105-116-102-**115**-116-11-12-117-110-1-111-112-4-5-113-114-**8.**

S4:

8-22-7-21-6-20-5-19-4-18-3-17-2-16-1-15-14-28-13-27-12-26-11-25-10-24-9-**23**-24-25-26-27-28-15-29-16-17-18-19-20-21-22-**35.**

S5:

35-34-21-33-20-32-18-31-17-30-29-40-28-39-27-38-25-37-24-**36**-37-51-38-52-39-40-41-30-31-44-32-45-33-34-**48**-47-33-46-45-59-44-43-31-42-41-54-39-53-52-65-51-50-37-**49**-50-65-66-53-54-55-42-43-48-59-46-47-**62**-61-46-60-58-57-43-56-55-68-53-67-66-64-50-**63**-64-67-68-56-57-60-61-**69.**

S6:

69-57-67-69-(57-67-69).

Return:

69-63-62-49-48-36-35-23-*8*-115-99-100-78.

The complete walk is sliced into the steps of the BNM. The last vertex traversed in each step is the place where the next step begins. Therefore, though the vertex is counted twice, practically it is not the repetition of the vertex. Traversing the triangle in S6 once symbolizes that the dancer has reached very close to the deity and taken a round around it (pradakShina). She traverses the triangle once again symbolizing that she reaches the deity and offers a pradakshina to it. This apparent repetition is shown in the bracket. The portion at the tail-end of the walk, labeled *return*, marks the descent back to the starting point (vertex 78). The vertex 8, italicized, indicates the point where the dancer returns to the outer circles.

Defining a Walk in SCG to Map it BNM

The walk in the EC of SGC changes direction from clockwise to anticlockwise each time it enters a new layer (The vertices at which this change takes place are shown in bold). Therefore, while a dancer following it begins by orbiting in a clockwise direction, throughout the dance half the time she would be moving in the anticlockwise direction around the center, but that is not acceptable to the tradition. Also, according to the general convention of the ritual of worship among Hindus, a devotee has to walk around a deity in complete rounds. Leaving an orbit incomplete, entering into the next one and completing all incomplete orbits in the end, one-by-one, as it is happening in the EC, is also not acceptable to the tradition.

Recall that the 9th Avarana is the world in which the performance is taking place. The dancer enters into the 8th Avarana through Bhoopuram, say at vertex 78. There onwards, an Avarana is an assembly of petals or triangles. For brevity let us consider that Bhoopuram has four petals pointing towards the four corners of the dance-stage. Petals or triangles are structures characterized by 3 points, 2 at the base and one at the apex, that are to be identified with respectively the two knees and head of a BN dancer in the Mandali posture. By this logic, the midpoint of the base of a petal or triangle is the location where the dancer takes a stance.

Practically speaking, the on-stage transitions will be equivalent to a walk joining the midpoints of the base of the petal (or triangle). So, kritis or gestures in each Awarana are to be choreographed for a dancer walking along its base. The dancer's transition to the next Avarana is recommended at a petal (or triangle) edge that is incident on the vertex with which her left knee is identified. It will take her a bit ahead of the point on S where she has to reach by walking a small distance anticlockwise. This small distance will be traversed clockwise as soon as she begins to perform in the Avarana. This arrangement leads to a vertical chain of left halves of the petals and triangles below the center during the ascent of the dance, while the corresponding right halves are generated during the descent (Fig. 7).

The BNM walk in Fig. 7 (the portions in which an edge is repeated by traversing it back immediately are marked in underlined): 78-79-80-81-82-83-84-85-70-71-72-73-74-75-76-77-78-100-99-100-101-102-103-104-105-106-107-108-109-86-87-88-89-90-91-92-93-94-95-96-97-98-99-115-8-115-116-11-12-117-110-1-111-112-4-5-113-114-8-23-35-23-24-25-26-27-28-15-29-16-17-18-19-20-21-22-35-36-48-36-37-51-38-57-39-40-41-30-31-44-32-45-33-34-48-49-62-49-50-65-66-53-54-55-42-43-58-59-46-47-62-63-69-63-64-67-68-56-57-60-61-69-67-57-69-69-67-57-69-63-64-67-68-56-57-60-61-69-61-62-49-50-65-66-53-54-55-42-43-58-59-46-47-62-47-48-36-37-51-38-52-39-40-41-30-31-44-32-45-33-34-48-34-35-36-37-51-38-52-39-40-41-30-31-44-32-33-34-35-23-24-25-26-27-28-15-29-16-17-18-19-20-21-22-35-22-8-23-24-25-26-27-28-15-29-16-17-18-19-20-21-22-35-22-8-117-118-119-11-12-120-121-110-1-111-112-113-4-5-114-115-116-8-116-99-98-100-101-102-103-104-105-106-107-108-109-86-87-88-89-90-91-92-93-94-95-96-97-98-99-98-78-79-80-81-82-83-84-85-70-71-72-73-74-75-76-77-78.

This walk above leads to the completion of seven orbits in the ascent and the same seven orbits in reverse order during the descent. Due to a requirement of BNM, there is an additional orbit inside the Trikona chakra.

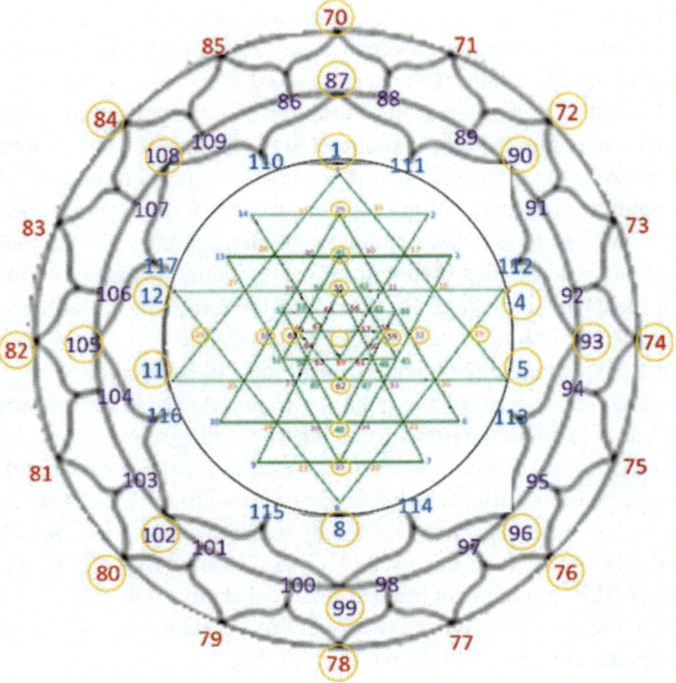

Fig. 7 BNM walk in SCG

Confirmation of our Logic of Mapping BNM to SGC

Assuming that a BN performance of SC dance is roughly 90 min long, the tentative number of minutes spent at each step (Ti, in seconds) and tempo of the music in it are generated according to the suggestions of domain experts. Recall from algorithm DrawSC that the SC design is inscribed in a circle of radius 48 units. The circumference of the base circle of each Avarana has been employed to compute a rough estimate of the length to be traversed in each of the steps. The distance to be traversed per second (Li/Ti) has been computed (seeTable 4).

The information provided by the traditional dance authorities and our estimates are placed side by side in Table 4, with the rows being in descending order of the velocity of traversal (the distance per unit time to be traversed in the corresponding step). The higher this ratio, the faster the dancer has to move. The traditionally recommended tempos are consistent with the ones those are indicated by our computed figures. Arguably, this consistency could be considered as a qualitative validation of our approach.

Further, the velocities of traversal suggest that a devotee would be able to complete as many as six orbits along the Trikona Chakra at speeds lower than the one attained in the Varnam. This gives the dancer an opportunity of walking around the Deity 5 to

Table 4 For the comparison of tempo and distance to be traversed per second

Step No	Step in the BNM	AvaraNas traversed	Tempo	Time (Ti)	Length (Li)	Li/Ti
7	JAvali	CAT and 8-Petals	Fast	480	482.87	1.006
8	TillAnA	16-Petals and Bhoopuram	Fast and medium	540	514.31	0.952
1	Invocations	Bhoopuram	Slow, medium and fast	300	276.32	0.921
2	Alarippu	16-Petals	Medium	300	237.99	0.793
3	Shabdam	8-Petals	Medium	600	174.62	0.291
4	JathiSwaram	14-Triangles	Medium	480	129.75	0.270
5	Varnam	Bahirdasha, Antardasha, 8-triangles and Trikona chakra	Slow and medium	1920	187.5	0.098
6	Padam	Trikona chakra	Slow	600	12.6	0.021
9	Shloka	Gopuram (Entry gate, source point of the SC dance)	Slow	300	1	0.003

10 times in close proximity (less than half the distance of the deity from Bhoopuram). The distance at which the devotee stands from the deity has a special importance in the worship.

The vertices circled in golden color are locations where the choreographer can design sequences of steps. The distance between two such vertices gives a clue for deciding on the steps. The smaller the distance, the simpler are the steps to be designed. This is consistent with the relatively simple steps in the initial parts of the BNM, reaching the climax at the Padam, and creating complex and attractive gestures while reaching the innermost circle. The number of estimated gestures in each circle are, respectively, 8, 8, 6 or 8, 4, 4, 4 and 1. These are to be decided according to the number of beats within the rhythmic cycle—no rhythm, 3 + 4 beats, 3 or 4 beats and 7 beats per cycle.

This study of SC drawing motivated us to think of a few more possibilities in the context of dance in general. These are compiled in the Section below. They call for further investigations.

Further Possibilities

1. Three variants of SC have been reported in the literature [15]. Polygons get inscribed in their bounding circles if the points touching the circle are joined by

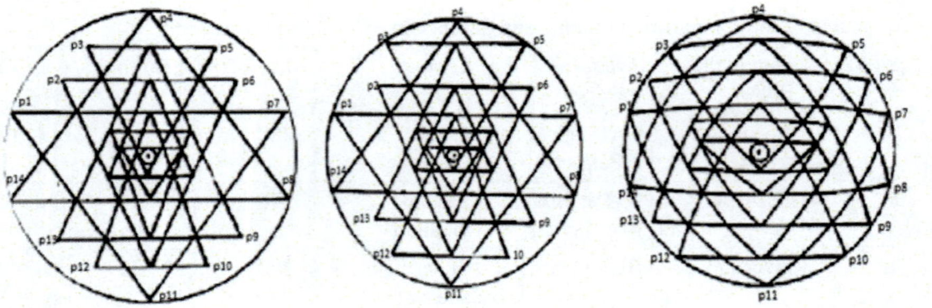

(a) (b) (c)

Fig. 8 Variants of SC; *Source* [15]

straight lines. There are 6, 8 and 14 such points in Fig. 8a–c, and accordingly, the bounding polygons will be hexagon, octagon and tetradecagon respectively.

These three variants might mean different things to a devotee or they might have differing appeal to different choreographers. Given that a computer-assisted drawing algorithm is ready, any of the drawings could be generated by tweaking the algorithm. For example, if we change the length of CS5 from 6 to 11.6 units in DrawSC, Fig. 9a changes to Fig. 9b, a more commonly seen variant of SC.

2. Solving the system of linear equations that models a perfect SC might well be an intractable problem for a modern computer. The meeting point of three lines might generate, depending on the context an insignificant or very small triangle [15]. For example, the SC generated by our algorithm shows 8 small triangles having

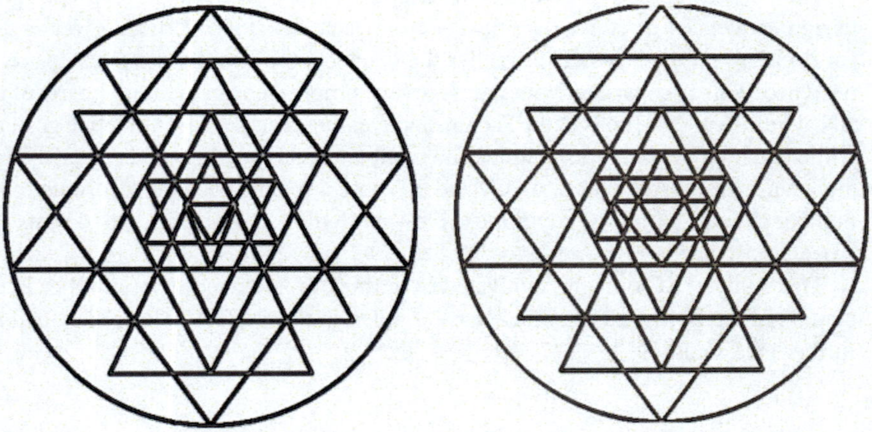

(a) SC with len(CS5) = 6 units (b) SC with len(CS5) = 11.6 units

Fig. 9 Outputs of DrawSC

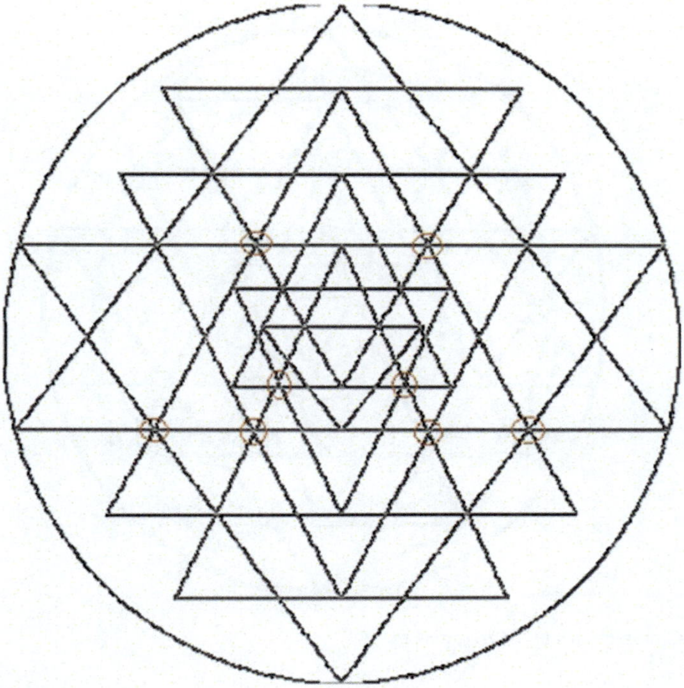

Fig. 10 .

areas less than 1/1000 of the area of the smallest visible triangle in the design (Fig. 10). Does it indicate human limitations in attaining perfection in worship? Could we explicitly attribute to this unavoidable imperfection the custom of not depicting the meeting of devotee and deity?

3. While SC dance as an illustration of BNM is for a solo performance, Fig. 9 (consisting of circumcircles drawn on the assembly of nine triangles in SC) clearly indicates the potential of the SC design for choreographing a group dance of 2 to 10 persons. A pair of circles marked in one color could be a clue for a pair-wise synchronized transition. Also, the situation that the circumcircle covering the line segment CS1 exceeds the limits of the bounding circle could be an indication of a small inherent error that would hinder the attainment of perfection in making all transitions perfectly circular and simultaneously fit the outermost circle. Or we speculate that it could be taken as a clue to display a strong bonding of a devotee to the material world that pulls her back while her soul is longing to meet the divinity (Fig. 11).
4. Generally, the Chakra images are aesthetically beautiful designs of varying complexity [15, p. 195]. Traditionally they are associated with one or the other deity [35]. Like SC, each of them could be a candidate for a dance. The methodology generated as a part of this study is able to provide Chakra designs fitting

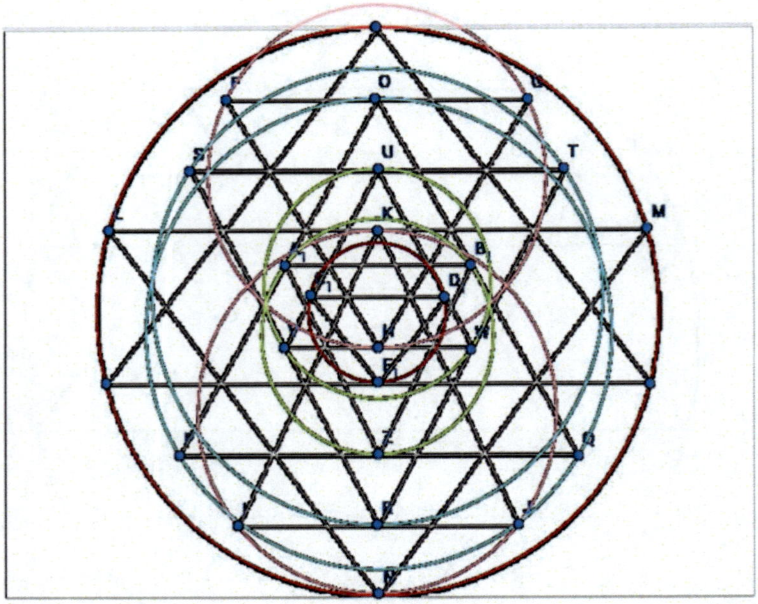

Fig. 11 Circumcircles of the triangles in CAT

the required dimensions. Then Chakra dance could be a new specialization for dance choreographers.

5. A different variant of the EC of SCG has been developed [15]. This path would facilitate a devotee to get to the deity quickly and allow more time for performing on the return journey, i.e., once the goal has been achieved. The system discussed in this research is capable of generating diverse mappings by following differing graph traversal algorithms. Choreographers may attach semantics to them, or the mappings could be generated after semantics has been provided.
6. Interactive editors for jotting down the choreographer's ideas are available [16]; customization would be required for the various classical dances of India. Elementary research in creating Adavus automatically has been presented [36]. This work could be extended to machine assisted creation of advanced choreography. The results reported in this paper are about setting up a path for the dancer's transitions on a stage. Information is available about other parameters, like rhythm and tempo at each step, the semantics, and choice of gestures matching the design [37]. Designing an expert system for automatic creation of a dance is a research problem. Such a system may help contribute to the preservation, tutoring and research of dance.

Conclusion

The topic of this paper is a systematically developed interdisciplinary research initiative to equip the choreographers and dancers of a famous Indian classical dance. Machine-generated maps of the dance floor are created, with annotations of clues for the expected transitions. Pointers are presented to a knowledge base for assisting dance professionals and for the development of the discipline. A few assertions and their proofs as well as experiments and their results have been discussed to demonstrate the key concept of mapping BNM to SC design.

The illustration of BNM with SC dance as a case study has been documented for the first time in the history of BN. This may inspire dance composers to choreograph it; this has been a demand over the years. Attempting such a choreography true to the actual design of SC has been a big challenge for a human choreographer, due to the complexity of the design and the number of circularly woven interlacing triangles, their relative locations and length measurements. Plotting SC to fit the size of a given stage or meet the requirement of a choreographer is a challenge that has been overcome by providing an automated system. A graph theoretic perspective of SC has opened up a new avenue for interdisciplinary research. A mapping of BNM to SC is one of the important achievements of the present research. The possibilities of enriching the dance domain by seeking assistance from computer-generated information have been demonstrated; the proposed SC group dance and the concept of Chakra dance as a specialization are two examples.

References

1. M. Joshi, S. Jadhav, An Extensive Review of Computational Dance Automation Techniques and Applications (2019), arXiv:1906.00606v1. [cs.HC]
2. K. Shinozaki, A. Iwatani, R. Nakatsu, Concept and construction of a dance robot system, in *Proceedings of 2nd International Conference on Digital Interactive Media in Entertainment and Arts* (2007)
3. D. Hariharan, T. Acharya, S. Mitra, Recognizing hand gestures of a dancer, in *Proceedings of International Conference on Pattern Recognition and Machine Learning* (2011)
4. S. Paul, S.N. Sinha, A. Mukerjee, Virtual Kathakali: Gesture-Driven Metamorphosis (1998), arXiv:cs/9812023v1 [cs.HC]
5. R. Majumdar, P. Dinesan, Framework for teaching Bharatanatyam through digital medium, in *Proceedings of 4th International Conference on Technology for Education* (2012)
6. S. Saha, L. Ghosh, R. Janarathanan, Gesture Recognition for Bharatanatyam Dance, in *Proceedings of 5th International Conference on Computational Intelligence and Communication Networks* (2013)
7. K.E. SukeL, R. Catrambone, G. Brostow, Presenting movement in a computer-based dance tutor. Int. J. Hum. Comput. Interact. (2003)
8. G. Dubbin, K. Stanley, Learning to dance through interactive evolution, in *Proceedings of Applications of Evolutionary Computation* (2010)
9. A. Curtis, J. Shim, E. Gargas, A. Srinivasan, A.M. Howard, Dance dance pleo: developing a low-cost learning robotic dance therapy aid, in *Proceedings of the 10th International Conference on Interaction Design and Children* (2011)

10. B. Rao, K. Rajni, The many facets of muthuswami dikshitar. J. Indian Musicol. Soc. **6**(3) (1975)
11. N. Altshiller-Court, *College Geometry: An Introduction to the Modern Geometry of the Triangle and the Circle*, 2nd edn. (Barnes & Noble, New York, 1925)
12. H.E. Huntley, *The Divine Proportion: A Study in Mathematical Beauty* (Dover Publication Inc., 1970)
13. M. Majewski, Geometric Ornament in Art and Architecture of Western Cultures, [Online]. Available: http://atcm.mathandtech.org/ep2012/invited_papers/3472012_19700.pdf. Accessed 6 Dec 2017
14. A. Nastasi, 10 Amazing Examples of Architecture Inspired by Mathematics (20 September 2012). [Online]. Available: http://flavorwire.com/330293/10-amazing-examples-of-architecture-inspired-by-mathematics/10. Accessed 6 Dec 2017
15. P.M. Sindhu, Information encoding and network security applications of Yantra images from manuscripts. Ph.D. Thesis, University of Mumbai, Mumbai, 2020.
16. G. Savage, J. Officer, CHOREO: an interactive computer model for dance. Int. J. Hum. Comput. (Formerly, Man-Machine) Stud. (1978)
17. L. Wilke, T. Calvert, R. Ryman, I. Fox, From dance notation to human animation: the Laban dancer project. Comput. Animat. Virtual Worlds **16**(3–4), 201–211 (2005)
18. D. Grunberg, R. Ellenberg et al., Development of an autonomous dancing robot. Int. J. Hybrid Inf. Technol. (2010)
19. A. Nakazawa, S. Nakaoka, K. Ikeuchi, Digital archive of human dance motions, in *Proceedings of the International Conference on Virtual Systems and Multimedia* (2002)
20. S. Jadhav, M. Joshi, J. Pawar, Art to SMart: an automated Bharatanatyam dance choreography. Appl. Artif. Intell. (2015)
21. F. Lapointe, Choreogenetics: the generation of choreographic variants through genetic mutations and selection, in *Proceedings of 7th Annual Workshop on Genetic and Evolutionary Computation* (2005)
22. N. Gwee, Two sub-optimal algorithms for an NP-hard dance choreography problem comparison of genetic and greedy process implementation. GSTF J. Comput. 52–58 (2015)
23. T. Mallick, P. Das, A. Majumdar, Posture and Sequence Recognition for Bharatanatyam Dance Performances using Machine Learning Approach (2019), https://doi.org/10.48550/arXiv.1909.11023 [cs.CV]
24. A. Mallik, S. Chaudhury, H. Ghosh, Preservation of intangible heritage: a case study of Indian classical dance, in *Proceedings of the Second Workshop on eHeritage and Digital Art Preservation* (2010)
25. V. Mamania, A. Shaji, S. Chandran, Markerless motion capture from monocular videos, in *Proceedings of the Indian Conference on Computer Vision, Graphics and Image Processing* (2004)
26. I. Hagendoorn, Emergent patterns in dance improvisation and choreography, in *Unifying Themes in Complex Systems IV* (2008)
27. M. Nakazawa, A. Paezold-Ruehl, DANCING, dance and choreography: an intelligent nondeterministic generator, in *Proceedings of the 5th Richard Tapia Celebration of Diversity in Computing Conference: Intellect, Initiatives, Insight and Innovations* (2009)
28. J. Stuart, E. Bradley, Learning the Grammar of dance, in *Proceedings of the 15th International Conference on Machine Learning* (1998)
29. D. Conrad, Photo, artwork by Daniel Conrad (Chromacon); Directed by Harish Johari, [Online]. Available: https://commons.wikimedia.org/w/index.php?curid=17021512
30. P. Sivadatta, Kasinath, P. Parab, T. Javaji, *The Natya Shastra of Bharata Muni* (Nirnaya Sagara Press, Bombay, 1894)
31. Chakra, Wikipedia, [Online]. Available: https://en.wikipedia.org/wiki/Chakra
32. Pancha Bhoota, Wikipedia, [Online]. Available: https://en.wikipedia.org/wiki/Pancha_Bhoota
33. R.J. Wilson, *Introduction to Graph Theory*, 4 edn. (Addison Wesley Longman Limited, 1996)
34. R.A. Dunlap, *The Golden Ratio and Fibonacci Numbers* (World Scientific, New Jersey, 1997)
35. M. Brand, A. Hertzmann, Style machines, in *Proceedings of 27th on Annual Conference on Computer Graphics and Interactive Techniques* (2000)

36. S.N. Kanippayyur, *Yantra Vidhikal (Malayalam), Kunnamkulam* (Panchangam Press, Kerala, 2007)
37. T. Mallick, P. Das, A. Majumdar, *Characterization, Detection, and Synchronization of Audio-Video Events in Bharatanatyam Adavus* (Springer, Heritage Preservation, 2018)
38. P. Suresh, *Natya Tantra—The Liberating Dance* (Aatmalaya Academy of Art and Culture Trust, Bangalore, 2016)

Computation of 22 Shrutis: A Fibonacci Sequence-Based Approach

Ambuja Salgaonkar

Preamble: The microtonal 22-shruti system proposes the tones, the atomic elements, and thereby lends a few clues for developing a basic structure for the compositions in Hindustani classical music (HCM). It has been understood that the consonance and dissonance of the notes that are based upon this shruti system leads to the creation of moods through the raagas, the melodic framework of HCM [7]. The known oldest reference to the 22-shruti system is in Natyashaastra, a Sanskrit treatise on performing arts, dated to between 200 BCE to 200 CE [1]. Over the time though the music has evolved along with the taste of its appreciators, the shruti computations remained a topic of curiosity for the musicians, mathematicians and engineers for centuries [6, 9]. Efforts have been made to produce instruments to demonstrate the 22 shrutis and to employ such instruments in order to complement the vocal music as closely as possible [5, 6]. Have the shrutis been invariant over the time is a question to be answered.

There are multiple issues. Computation of 22-shrutis is a problem of reverse engineering. That means any method that produces a set of acceptable 22- shrutis is acceptable. The diverse methods that have been suggested [4, 6, 9], could be compared in terms of the complexities of the algorithms and axioms they follow. Apparently, the shruti values computed and adopted by genius musicians are not identical [6]. This casts a doubt on the reliability of the entity itself. There is a possibility of apparent diversity and internal consistency of these sets that need to be researched. The recent publications discuss what they claim as a robust interpretation. It follows the axiom of the two consonances, i.e., with the fifth and the third notes respectively, with respect to the aadhaar Shad-ja, literally meaning baseline note, that gives birth to the remaining 6 pure notes [17] of HCM.

A. Salgaonkar (✉)
Department of Computer Science, University of Mumbai, Mumbai 400098, India
e-mail: ambujas@udcs.mu.ac.in

A. Salgaonkar and M. Velankar (eds.), *Computer Assisted Music and Dramatics*,
Advances in Intelligent Systems and Computing 1444,
https://doi.org/10.1007/978-981-99-0887-5_13

Musicians have a question, even if we could mathematically resolve such issues, how would it contribute to enriching the music? Do the practitioners actually employ this implicit knowledge is yet another, probably the most important, question from the application point of view.

This paper aims at answering some of these questions and providing pointers to address a bigger question that is of automatic creation of valid structures in HCM. Specifically, we consider [9] as a basis for investigations presented here.

First, we give basic terminology. Followed by it are the details of our findings.

Terminology and Shruti Computations Using Swaymbhu Gandhaar

The musical *notes* are treated here as singular frequencies, taking only the fundamental frequency rather than the entire timbral complex of each note into account. Ratios of relative frequencies define relative positioning of notes in HCM. For example, the notes that are $(2)^k$ apart are identified with the same name. There are 7 notes, Shad-ja (sa), Rishabha (re), Gandhaar (ga), Madhyama (ma), Panchama (pa), Dhaivata (dha) and Nishaada (ni) in the octave of HCM. The next to ni comes sa of the next octave. They are considered *shuddha* swara or pure notes. They roughly correspond to the western major scale (WS). re, ga, dha and ni have variants called *komal*, roughly corresponding to the flat notes in WS, while the variant of ma is *teevra* that corresponds to sharp. The sa, the frequency ratio 1/1, could be identified with any of the keys of a piano and not necessarily to A or C as in western music. A set of 5 variants is obtained by employing *sa-pa bhaava*, the consonance with the fifth note (3/2) in ascending order, and the other by employing it in descending order, called *sa-ma bhaava*. The literal meaning of *bhaava* is nature or characteristic. Yet another set of 10 variants exist, that could be obtained by manipulating the 10 in hand with respect to 5/4, a special variant of the third note, called *swayambhu gandhaar*, literally means *gandhaar* that evolves naturally. It gives rise to *sa-ga bhaava*. No variants of sa and pa are accepted. Together there are $(5*2)*2 + 2 = 22$ notes, called shrutis. Details of these computations are given in Table 1.

Note that in the computations shown in Table 1, without loss of generality, we assume *madhya saptaka* or middle octave that begins at 1/1. The shruti values in Table 1 are maintained to be in the madhya saptaka by employing multiplication and division by 2.

The sa-pa and sa-ga consonances are axioms. They are not to be proved. It is like someone performed 3/2 shruti in coordination with 1/1 and it was found to be nice. So, they played 3/2 of the new one and that was also found to be nice. How long would it have continued? Is it as long as we find a new shruti?

Clearly, at third and fifth iterations the computations have reached the new octave. Therefore, it has taken three octaves to get the first of the 4 sets in Table 1. If the computed shrutis do not lie in the madhya saptaka then we have considered their

Table 1 Computation of 22 shruti

	Part I Iter_1(1) = 1; Iter_1(i + 1) = Iter_1(i) * 3/2; if Iter_1(i) > 2 then Iter_1(i) = Iter_1(i) / 2			Part II Iter_2(1) = 1; Iter_2(i + 1) = Iter_2(i) * 2/3; if Iter_2(i) < 1 then Iter_2(i) = Iter_2(i) * 2		
Iter#	Iter_1(i)	Shruti	Shruti#	Iter_2(i)	Shruti	Shruti#
1	1/1	sa	0	1/1	sa	0
2	3/2	pa	13	2/3 = 4/3	ma	9
3	**9/4 = 9/8**	**re**	**4**	**8/9 = 16/9**	**ni**	**18**
4	27/16	dha	17	32/27	ga	5
5	**81/32 = 81/64**	**ga**	**8**	**64/81 = 128/81**	**dha**	**14**
6	243/128	ni	21	256/243	re	1
7	729/256 = 729/512	ma	12			
	Part III For i: 3 to 7 Iter_3(i + 6) = Iter_3(i) * 80/81; if Iter_3(i) > 2 then Iter_3(i) = Iter_3(i) / 2			Part IV For i: 2 to 6 Iter_4(i + 6) = Iter_4(i) * 81/80; if Iter_4(i) < 1 then Iter_4(i) = Iter_4(i) * 2		
Iter#	Iter_1(i)	Shruti	Shruti#	Iter_2(i)	Shruti	Shruti#
8				27/20	ma	10
9	10/9	re	3	9/10 = 9/5	ni	19
10	5/3	dha	16	3/5 = 6/5	ga	6
11	5/4	ga	7	8/5	dha	15
12	15/8	ni	20	16/15	re	2
13	45/16 = 45/32	ma	11			

images from it. So, we could continue with the iteration logic as long as we wish or till there is no recurrent pattern observed in the output. It would give rise to a divergent series. The pattern won't repeat. However, [4, 9] and most of the modern musicians recommend to stop with the seven cycles when the first set of 7 shrutis is ready and compute the first variants of all but sa and pa by employing the sa-pa bhaava in the reverse direction i.e., by changing the multiplying factor to the reciprocal of 3/2. As the fifth note from sa in this direction is ma, the resultant is the set of 5 shrutis as obtained in part II, together with the *achala*, i.e., invariant, sa and pa, we get a saptaka in *sa-ma bhaava*.

One of the reasons for why the iterations with the multiplier 2/3 are dropped after the 7th is that as the iterations increase, the ratio becomes complex. Continuing with the same reason it has been stated that for many musicians the distance 5/4 (=1.25) being expressed as a ratio of small relative primes (only 1 is the common factor between them), is a "natural" choice over the 81/64 (=1.267) that is obtained for the third note of a saptaka with the sa-pa bhaava [4, 7]. Note that (5/4) * x = 81/64 → x

= (81/64) * (4/5) = 81/80. Therefore, 81/80 is the distance of swayambhu gandhaar from the gandhaar of sa-pa bhaava saptaka.

The ratio 81/80 is known as the *syntonic comma*, the adjustment that helps synchronize a tonic, that is, the first and the last note of a saptaka. One may identify a new sa with 80/81.

Acceptance to swayambhu gandhaar gives rise to the saptaka in the third part of Table 1. The one in part four is obtained by employing sa-ma with the swayambhu gandhar.

Several methods have been recommended for the computation of 22 shrutis [2, 4, 6, 7, 9]. Table 1 employs one of them. Hardly any of them answers why the syntonic comma is chosen to be 81/80 except the subjective explanation that it is the distance between two notes that any untrained ear could also notice and feel that these two notes are out of tune versions of a single note [18]. Clearly, there is a scope for investigation in the epistemology of the 22 shruti system.

The Shruti# beginning with 0 are the indices of the shrutis when all 22 shrutis of a specific saptaka are ordered ascendingly. Shruti# is independent of the order in which the shrutis are computed. For instance, 7 shrutis in a saptaka could be generated in the following manner as well: Multiply sa (1/1) by 3/2 → get pa; Multiply sa (2/1) by 2/3 → get ma; Compute ma-pa = distance (ma, pa) → identify re from sa and dha from pa at the distance ma-pa in one direction and ni from sa and ga from ma at the pa-ma distance in the other direction. Since the resultant shrutis remains unchanged, the shruti# will not change.

Summation of shruti distances is by the multiplication of their ratios. For example, distance(sa, re) + distance(sa, ga) = (9/8) * (81/64) = 729/512 = distance(sa, ma). Equivalently, the eighth shruti added to the fourth results in the twelfth shruti. Among the four shrutis of ma, the one obtained in this computation is the longest distance apart from sa. It is known as teevra (sharp) ma.

It is said that the whole system of computation of 22 shrutis is a game of playing with the ratios of small relative primes: 1/1, 2/1, 3/2 and 5/4 [9]. Even if we accept it, the question arises why to stop with 5/4? Why not choose 7/5 as the next candidate? In fact, why only 7 points itself is a question to be answered logically.

Research is called for the analysis of discrepancies in the known theory and practices, and to reason out apparent deviations and inconsistencies that have been labeled as accident or creativity [3]. Generating large annotated samples and employing machine learning to discover latent relationships is one way to attempt it. However, the annotation involves tremendous efforts by the learned musicians and the job is more complicated when they usually are not techno-friendly. On the other hand, with the advent of technology, visualization of a mathematical model that answers most of the questions in a consistent manner is a possible alternative. If such a formalization works satisfactorily then one may map it to several established theories and more knowledge could be built up. The work presented in this paper is of the latter type. Some of the readers may find the approach and the findings unusual. The author admits that they are. We hope that this research would serve as a basis for further research, like for the creation of the simulated databases, for musical profiling of the well-known artists and most importantly, for the automatic validation and creation of musical pieces.

In the next section we discuss four of the above issues in the form of our contentions and provide mathematical solutions to support them.

Contentions with Mathematical Back Up

C1: Only the axiom of sa-pa bhaava is sufficient for generating the 22 shrutis.

We propose the following logic

Block 1, 2, 4 and 7 of Table 2 gives a step to generate a set of 5 values, each consisting of single variants of re, ga, ma, dha and ni. The computations in the remaining blocks show that no ratios other than these are generated for these 5 notes.

Essentially, we have computed a new shruti [square 1 and 4] by adding a sa-pa distance in the one at hand. In other words, this distance has been considered as a fundamental unit for identifying a new shruti from a given location. Therefore, it needs to be maintained constant throughout the process. For this, given that sa is the baseline note, the tonic, and hence is fixed, pa has to be necessarily fixed with respect to it. Hence there are no variants of sa and pa.

Where did the ratio 16/15 come in? It could be verified that starting with 2/1 and continuing with 3/2, 4/3 and so on, 16/15 is the nearest ratio to 2187/2048. The distance between the two is 0.012. Therefore 16/15 is chosen to represent 2187/2048 as its close approximation.

Thus, we computed the unique set of 22 shrutis as in Table 1 only by employing the axiom of sa-pa bhaava. Swayambhu gandhaar is a product of it. During the process, multiple syntonic commas are generated though the status is given to only one of them. In this process, the exclusion of sa and pa from the shifts is not enforced but automatic. Also, the process has explained the natural bound on the number of iterations.

Where from the sa-pa bhaava came into being is the question to be answered yet. One of the answers is 2/1, 3/2, 4/3 and 5/4 are the best consonances among them 3/2 is the first and the most powerful after the 2/1 [9]. We deal with this question in the discussion on our 4th contention. Now we address a more allied question, is it the only authentic set of 22 shrutis since the ancient days of HCM?

C2: The above discussed set of 22 shrutis and the shrutis reported by several other musicians have been devised around the equal tempered scale.

The modern musicians' claim of a set of unique 22 shrutis follows the basis of Bilawal scale which is of all 7 pure notes that consisted of shruti# 0, 4, 7, 9, 13, 17 and 20. Its deviation from the scale obtained by employing pure sa-pa bhaava is the key to generate the 22 shrutis (Table 1). It is understood that Bilawal has been considered as a standard for HCM since the early nineteenth century [10]. This gives a clue for us to speculate that these may not be the same 22 shrutis described in ancient texts.

Also, the discrepancy in the figures reported for 22-shrutis by the various musicians calls for further investigations.

Like Rishi Panini formulated the grammar of modern Sanskrit from the corpus available at his time, historically, it was in the first decade of the twentieth century, Pandit V N Bhatkhande wrote what has now become the canonical treaties of the modern HCM [2] out of his relentless efforts in meeting practitioners and collecting

Table 2 Exhaustive listing of the steps for generating 22 shrutis only by employing sa-pa bhaava

Square 1	Square 2
sa = (1/1) → (1/1) * (3/2) = 3/2 — pa → (3/2) * (3/2) = 9/4 = 9/8 — re → (9/8) * (3/2) = 27/16 — dha → (27/16) * (3/2) = 81/32 = 81/64 — ga → (81/64) * (3/2) = 243/128 — ni → (243/128) * (3/2) = 729/256 = 729/512 — ma → _(729/512) * (3/2) = 2187/1024 = 2187/2048 — sa?_ Bring it back to 1/1	→ (2187/2048) * (2048/2187) = 1/1 — sa → (729/512) * (2048/2187) = 4/3 — ma → (243/128) * (2048/2187) = 16/9 — ni → (81/64) * (2048/2187) = 32/27 — ga → (27/16) * (2048/2187) = 128/81 — dha → (9/8) * (2048/2187) = 256/243 — re → _(3/2) * (2048/2187) = 1024/729 — pa?_ This is not pa but a new shruti! Therefore multiply each of these by 729/1024
Square 3	Square 4 /* (2187/2048) / (16/15) = 1.00113; Continue sa-pa computations with it. */
→ (1024/729) * (729/1024) = (1/1) — sa → (256/243) * (729/1024) = 3/4 = 3/2 — pa → (128/81) * (729/1024) = 9/8 — re → (32/27) * (729/1024) = 27/32 = 27/16 — dha → (16/9) * (729/1024) = 81/64 — ga → (4/3) * (729/1024) = 243/256 = 243/128 — ni → (1/1) * (729/1024) = 729/1024 = 729/512 — ma /* All ratios are exactly as in Part I of Table 1. No shift is required. */	16/15 = re → (16/15) * (3/2) = 8/5 — dha → (8/5) * (3/2) = 12/5 = 6/5 — ga → (6/5) * (3/2) = 9/5 — ni → (9/5) * (3/2) = 27/10 = 27/20 — ma → _(27/20) * (3/2) = 81/40 = 81/80 — sa?_ Bring it back to 1/1
Square 5	Square 6
→ (81/80) * (80/81) = 1/1 — sa → (27/20) * (80/81) = 4/3 — ma → (9/5) * (80/81) = 16/9 — ni → (6/5) * (80/81) = 32/27 — ga → (8/5) * (80/81) = 128/81 — dha → (16/15) * (80/81) = 256/243 — re /*All values are identical with that in Square 2*/ _And, (1/1) * (80/81) = 80/81 – pa?_ This is not pa but a new shruti! Therefore multiply each of these by 81/80	→ (80*81) * (81/80) = 1/1 — sa → (256/243) * (81/80) = 16/15 — re → (128/81) * (81/80) = 8/5 — dha → (32/27) * (81/80) = 6/5 — ga → (16/9) * (81/80) = 9/5 — ni → (4/3) * (81/80) = 27/20 — ma /*All values are identical with that in Square 5*/ and, _(1/1) * (81/80) = 81/80 — sa?_ /*All values are identical with that in Square 4. Thus Square 4 and 5 generate each other.*/

(continued)

Table 2 (continued)

Square 7 /*Multiply the set in Square 1 by 80/81*/	Square 8
→ (2187/2048) * (80/81) = 135/128 → (729/512) * (80/81) = 45/32 — ma → (243/128) * (80/81) = 15/8 — ni → (81/64) * (80/81) = 5/4 — ga → (27/16) * (80/81) = 5/3 — dha → (9/8) * (80/81) = 10/9 — re And, *(3/2) * (80/81) = 40/27— pa?* This is not pa but a new shruti! Therefore, multiply each of these by 27/40	→ (40/27) * (27/40) = 1/1 — sa → (10/9) * (27/40) = 3/4 = 3/2 — pa → (5/3) * (27/40) = 9/8 — re → (5/4) * (27/40) = 27/32 = 27/16 — dha → (15/8) * (27/40) = 81/64 — ga → (45/32) * (27/40) = 243/256 = 243/128 — ni → (135/128) * (27/40) = 729/1024 = 729/512 — ma /* All ratios are exactly as in Part I of Table 1.; No shift is called for. Also, 80/81 itself is a ratio of small relative primes. Therefore, further identifying a ratio of small relative primes that approximates it, is invalid. The process terminates */

the subject related information by traveling pan-India and by studying the Sanskrit granthas on music and performing arts. He created a structure for HCM by contextualizing the music-practices of his time with respect to the old framework. It is understood that Pandit Bhatkhande articulated the present formalization of the modern HCM in response to the colonial encounter and British orientalist understandings of Hindustani music.

At around the same time, Simon Steve's work on the equal tempered scale was published in Europe. It is possible that the 22 shrutis of modern HCM were articulated in order to make them compliant with the musical tradition of the West as well. A glance at Table 1 will reveal that the shrutis generated through the pure sa-pa bhaava have expanded the scope of the notes in order to accommodate the piano-keys of equal tempered scale within them. For example, the key $G^{\flat}$ is at the middle of the first and second while the next key G is at the middle of the third and the fourth shruti of re. There is a possibility that the 22 shrutis in Table 1 are not the same as they were accepted to ancient Indian texts or this set didn't exist until there was any opportunity to compare the music of India with its counter-world in the west. The swayambhu gandhaar might be of Indian origin. Or as the name suggests, at least it should have its source in the environment around the musicians who employed it. In the light of [16], the latter seems more logical.

Here we attempt to benchmark the apparently diverse sets of shrutis presented by 6 respectable personalities in HCM, namely, Acharekar (A), Muley (M), Clements (C), Omkarnath (O), Brihaspati (B) and Ranade (R) with that of the set embedding the modern standard Bilawal scale (Tables 3 and 4). The last but first column of each sub-table consists of the shrutis converted from their Hz equivalents as reported in [6]. The second column presents its cent equivalent. The distance between successive shrutis in each set has been computed in the first column. The deviation from the set embedding Bilawal scale (BS) that is given in the first sub-table with the values of the major notes in bold, is in the last column of each sub-table. The mean and standard deviation of the deviation values are computed for each set and the tables are presented in the ascending order of the mean. Table 3 consists of the three tables with relatively lower mean values and from there it is clear that the values produced by the recent musicians are very close to the BS. While such a conclusion is difficult in the case of the three in Table 4. The smaller the value of the standard deviation, the greater the consistency of the model with respect to the BS.

The absolute values in each table are compared with the table of the BS and the rows are aligned accordingly. 0 (shown in bigger font) indicates that the values are identical. A bold 0 is suggestive of the consonance they might have employed. It has been obtained by analyzing the pattern of the distances between the successive keys. A pair of successive shrutis that embeds a key of an equal tempered scale piano or a shruti that is close to such a key is underlined or indicated in bold. The bold values in the first and last columns of each set indicate a discipline in the distribution of inter-shruti-distances. The underlined values in the same columns indicate that the pattern is locally disturbed. Cell with no color means nothing is revealed about its relation with shruti distribution pattern. When there are only 21 values presented, the missing value is supposedly the sa at 1/1. Shruti marked in bold and underlined are computed by the model.

From Tables 3 and 4 we can see that Bilawal-centric 22-shrutis being the analytically computed set have got a nice structure. Ranade (1937–2011) and Omkarnath (1897–1967) seem to have attempted to reach a similar structure on their own. However, Srijan Deshpande, a music scholar, pointed out that an artist does not compute shrutis while performing. Shrutis are natural to him/her, and may be the result of the performer's response to the complex timbral-acoustic environment, especially that generated by the tanpuras, in reference to which s/he produces them. Does it mean that though there is a mathematically confirmed standard available, there is a scope for subjectivity? Maybe the standards need to be a part of training and subjectivity is acceptable as it creates scope for an individual's style. Therefore, arguably, non-standards also need to be a part of training. Otherwise, by rejecting them the tradition may lose the diversity aspect of this art. Mr Srijan mentioned that Kumar Gandharva (1924–1992) and Abdul Karim Khan Saheb (1872–1937), both the stalwarts of HCM, are known for being accurate in shrutis. However, the musicians agree that even if they recite a particular raaga, say Todi, their shrutis and perspectives towards shrutis are different. Not only that, if only Kumar Gandharva's Todi is considered, different shrutis of re are observed at different places. An attempt to benchmark the shruti sets reported by say half a dozen authorities in music may generate some inputs for the musicians to reason out or interpret such diversities.

Knowing things analytically has the biggest advantage that most of the possibilities and impossibilities are known without doing any experimentation that is subjective or costly. Our next exploration is to find out the roots of sa-pa bhaava.

C3: The distance 3/2 is fundamental to HCM and then are the distances 5/3 and 8/5; not 5/4.

Table 3 Comparison of shrutis computed by Ranade, Omkarnath and Acharekar with BS

Dist_ Succ	Dist_ Cents	BS	Shruti	Dist_ Succ	Dist_ Cents	Ranade	Dev_ BS	Dist_ Succ	Dist_ Cents	Omkarnath	Dev_ BS	Dist_ Succ	Dist_ Cents	Acharekar	Dev_ BS
	0.00	**1**	**Sa**		0.00	1	**0**	**0.0125**	21.51	1.0125	0.0125	**0.0101**	17.41	1.01011	0.0101
0.0535	**90.23**	1.0535	re1	0.0534	**90.06**	1.0534	0.0001	**0.0416**	**92.04**	1.0546	0.001	**0.0429**	**90.06**	1.0534	0.0001
0.0125	**111.79**	1.0667	re2	0.0137	**113.57**	1.0678	0.001					**0.0125**	111.62	1.0666	0.0001
0.0416	182.39	1.1111	re3	0.0393	180.36	1.1098	0.0012	0.0489	174.73	1.1062	0.0044	**0.0535**	201.91	1.1237	0.0113
0.0125	203.91	**1.125**	re4	0.0137	203.91	1.125	0	0.0169	203.76	1.1249	0.0001	0.0125	223.34	1.1377	0.0113
								0.0125	225.32	1.139					
0.0535	**294.16**	1.1852	ga1	**0.0534**	**294.01**	1.1851	0.0001	**0.0417**	**296.05**	1.1865	0.0011	0.0417	**294.01**	1.1851	0.0001
0.0125	**315.64**	1.2	ga2	0.0126	**315.64**	1.2	0					0.0114	**313.62**	1.1986	0.0012
0.0417	386.31	**1.25**	ga3	0.0404	384.23	1.2485	0.0012	**0.0535**	386.31	1.25	0				
0.0125	407.79	1.2656	ga4	**0.0137**	407.79	1.2656	0	**0.0125**	407.79	1.2656	0	0.0546	405.73	1.2641	0.0012
								0.0125	429.27	1.2814					
0.0535	**498.00**	**1.3333**	**ma1**	0.0535	**498.00**	1.3333	**0**	**0.0417**	**499.95**	1.3348	0.0011	**0.0547**	**498.00**	1.3333	**0**
0.0125	**519.55**	1.35	ma2									**0.0113**	**517.50**	1.3484	0.0012
0.0417	590.29	1.4063	ma3	**0.0535**	588.19	1.4046	0.0012	**0.0535**	590.16	1.4062	0.0001	**0.0417**	588.19	1.4046	0.0012
0.0124	611.70	1.4238	**ma4**	**0.0137**	611.70	1.4238	0	**0.0125**	611.70	1.4238	**0**	**0.0125**	609.75	1.4222	0.0011
				0.0393	678.37	1.4797		0.0329	**667.80**	1.4707					
0.0535	**701.96**	**1.5**	**Pa**	**0.0137**	**701.96**	1.5	0	0.0326	**723.40**	1.5187	0.0125	**0.0547**	**701.96**	1.5	0
												0.0113	721.47	1.517	
0.0535	792.13	1.5802	dha1	**0.0535**	792.13	1.5802	0	0.0417	794.10	1.582	0.0011	**0.0417**	792.13	1.5802	0
0.0125	813.69	1.6	dha2	**0.0137**	815.74	1.6019	0.0012					**0.0125**	813.69	1.6	0
0.0417	**884.39**	1.6667	dha3	**0.0392**	**882.31**	1.6647	0.0012	**0.0535**	**884.29**	1.6666	0.0001	0.0534	**903.81**	1.6855	0.0113

(continued)

Table 3 (continued)

Dist_ Succ	Dist_ Cents	BS	Shruti	Dist_ Succ	Dist_ Cents	Ranade	Dev_ BS	Dist_ Succ	Dist_ Cents	Omkarnath	Dev_ BS	Dist_ Succ	Dist_ Cents	Acharekar	Dev_ BS
0.0125	**905.87**	**1.6875**	dha4	**0.0137**	**905.87**	1.6875	0	**0.0125**	**905.76**	1.6874	0.0001	0.0125	925.35	1.7066	0.0113
								0.0125	927.28	1.7085					
0.0535	996.11	1.7778	ni1	**0.0535**	996.01	1.7777	0.0001	**0.0417**	997.96	1.7797	0.0011	0.0417	996.01	1.7777	0.0001
0.0125	1017.6	1.8	ni2	**0.0137**	1019.5	1.802	0.0011					0.0114	1015.6	1.7979	0.0012
0.0417	**1088.3**	**1.875**	ni3	**0.0393**	**1086.2**	1.8728	0.0012	**0.0535**	**1088.3**	1.875	0				
0.0125	**1109.7**	1.8984	ni4	**0.0137**	**1109.7**	1.8984	0	**0.0125**	**1109.7**	1.8984	0	**0.0547**	**1107.8**	1.8962	0.0012
						Average	0.0004			Average	0.0016			Ave = rage	0.0029
						Std dev	0			Std dev	0.0002			Std dev	0.0002

Table 4 Comparison of shrutis computed by Brihaspati, Clements and Mule with BS

Dist_Succ	Dist_Cents	BS	Shruti	Dist_Succ	Dist_Cents	Brihaspati	Dev_BS	Dist_Succ	Dist_Cents	Clements	Dev_BS	Dist_Succ	Dist_Cents	Mule	Dev_BS
	0.00	**1**	**sa**					**0.0158**	27.13965488	1.0158	0.0158	0.0102	17.56913626	1.0102	0.0101
0.0535	**90.23**	1.0535	re1	**0.0528**	**89.07767302**	1.0528	0.0007	**0.0417**	**97.93478743**	1.0582	0.0045	0.0634	**106.4212458**	1.0634	**0.0093**
0.0125	**111.79**	1.0667	re2	**0.0131**	**111.6230798**	1.0666	0.0001	**0.0125**	**119.3966414**	1.0714	0.0044	0.1198	195.8893022	1.1198	**0.0474**
0.0416	182.39	1.1111	re3					**0.0535**	209.5944954	1.1287	0.0033	0.1343	218.1627062	1.1343	**0.0205**
0.0125	203.91	**1.125**	re4	**0.0532**	201.2919361	1.1233	0.0015	0.0125	231.0875297	1.1428		0.1485	239.7010281	1.1485	**0.0205**
				0.0128	223.3442204	1.1377									
0.0535	**294.16**	1.1852	ga1	**0.0135**	246.6211595	1.1531	**0.0271**	**0.0417**	**301.7357179**	1.1904	0.0044	0.1964	**310.4397787**	1.1964	0.0094
0.0125	**315.64**	1.2	ga2	**0.0407**	**315.641287**	1.2	**0**	**0.0125**	323.2707347	1.2053	0.0044	0.2604	400.6579914	1.2604	0.0479
0.0417	386.31	**1.25**	ga3	**0.0533**	405.5957562	1.264	**0.0112**	**0.0535**	413.5215396	1.2698	0.0158	0.2761	422.0896663	1.2761	0.0205
0.0125	407.79	1.2656	ga4	**0.0127**	427.3725723	1.28	0.0114	0.0091	429.2650747	1.2814	0.0125	0.2921	443.6612753	1.2921	0.0205
0.0535	**498.00**	**1.3333**	ma1	0.0125	**448.8788619**	1.296	**0.028**	0.0451	**505.6457196**	1.3392	0.0044	0.3459	**514.2854665**	1.3459	**0.0094**
0.0125	**519.55**	1.35	ma2	0.0417	**519.5512887**	1.35	0	0.0114	525.184654	1.3544	0.0033	0.4179	604.5069449	1.4179	**0.0479**
0.0417	590.29	1.4063	ma3									0.4356	625.9845939	1.4356	**0.0204**
0.0124	611.70	1.4238	ma4	0.0535	609.7492337	1.4222	**0.0011**	0.0547	617.4012435	1.4285	0.0033	0.4955	**696.7534925**	1.4955	0.0479
				0.0125	631.282574	1.44									
0.0535	**701.96**	**1.5**	**Pa**	0.0417	**701.9550009**	1.5	0	0.0501	**701.9550009**	1.5	**0**	0.5314	737.8213949	1.5314	**0.0205**
								0.0159	729.2082724	1.5238					
0.0535	792.13	1.5802	dha1	**0.0533**	791.9094701	1.58	0.0001	**0.0417**	799.8897883	1.5873	0.0045	0.5952	808.4847778	1.5952	**0.0094**
0.0125	813.69	1.6	dha2	**0.0127**	813.6862861	1.6	0	**0.0125**	821.3516423	1.6071	0.0044	0.6805	**898.6686512**	1.6805	**0.0479**
0.0417	**884.39**	1.6667	dha3									0.7015	920.1685816	1.7015	**0.0205**
0.0125	**905.87**	**1.6875**	dha4	**0.0532**	**903.4010505**	1.6851	**0.0014**	0.0518	**908.8376069**	1.6904	0.0017	0.7228	941.7062744	1.7228	**0.0205**

(continued)

Table 4 (continued)

Dist_Succ	Dist_Cents	BS	Shruti	Dist_Succ	Dist_Cents	Brihaspati	Dev_BS	Dist_Succ	Dist_Cents	Clements	Dev_BS	Dist_Succ	Dist_Cents	Mule	Dev_BS
				0.0128	925.3499438	1.7066		0.0091	924.4367118	1.7057					
0.0535	996.11	1.7778	ni1	0.0125	946.923861	1.728	**0.028**	0.005	933.0425305	1.7142	0.0358	0.7946	1012.39478	1.7946	**0.0094**
0.0125	1017.60	1.8	ni2	0.0417	1017.596288	1.8	0	0.0547	1025.273613	1.808	0.0044	0.8929	**1104.717836**	1.8929	**0.0491**
0.0417	**1088.27**	**1.875**	ni3									0.9142	1124.089889	1.9142	**0.0205**
0.0125	**1109.74**	1.8984	ni4	**0.0533**	**1107.550757**	1.896	**0.0013**	0.0535	**1115.476541**	1.9047	0.0033	0.9381	1145.561981	**1.9380**	**0.0205**
				0.0127	1129.327573	1.92									
						Ave rage	0.0051			Ave rage	0.0059			Ave rage	0.025
						Std dev	0.0004			Std dev	0.0004			Std dev	0.025

We are taught that the fundamental notes of our music were inspired by our Rishis, literally meaning the nature scientists, from their surroundings. Subscribing to this, we started exploring the roots of this fundamental ratio in nature. The perception that started at 1 by completing one octave we reach 2, completion of the next octave doesn't reach us 3, it reaches us 4 and likewise, is evidential that its topology has to be a more complex structure than a line or a circle. Marathi phrase "Suranchya sundar valayakruti" (beautiful curls of the notes) is common to describe the experience of improvisation. Also, by completing an octave we automatically do not reach the starting point but some distance ahead of it [9]. Together it gives us a clue that it may be a spiral topology projected or mapped to a circular one as explained in [4]. Aesthetics being a core property of these curls, the golden ratio [13] attracted our attention. From there we reach the golden spiral and its approximation [14, 15], the Fibonacci spiral. In the next paragraphs we describe the Fibonacci spiral and its application to the shruti cycle.

The Fibonacci numbers are defined by the recurrence relation: F(n) = F(n-1) + F(n-2) for n > 1 and F(1) = 1, F(0) = 0 [11]. The first 15 Fibonacci numbers are: 0, 1, 1, 2, 3, 5, 8, 13, 21, 34, 55, 89, 144, 233, 377, 610.

A 2-D Fibonacci rectangle tiling could be constructed as follows: S1: Draw a square of unit side. S2: Construct the next square on its right side, resulting in a rectangle of size 2 × 1. S3: Construct a square on the top of it, i.e., with side 2, resulting in a rectangle of size 3 × 2. S4: Construct a square on the left side, i.e., with side 3, resulting in a rectangle of size 5 × 3. S5: Construct a square on its bottom, i.e., with side 5, resulting in a rectangle of size 8 × 5.

Continue the construction in this manner, you will construct the squares with sides 8, 13, 21, 34 successively. Then with sides 55, 89, 144 and 233. So on, and so forth.

A Fibonacci spiral could be constructed by joining a pair of opposite corners of the first square by a smooth arc and then continuing the arc by joining the diagonally opposite corners of the next square and likewise. The resultant Fibonacci spiral is shown in Fig. 1 [12] below.

Clearly, length of a Fibonacci spiral is the sum of the lengths of its arcs passing through different squares and length of each of these pieces would be half of 3.14*side of the square. Note that the spiral goes through four consecutive squares to complete one turn. The partial spiral lengths at the end of each turn are computed for the first 5 turns: 10.99, 84.78, 590.32, 4055.31 and 27,804.7; Alternatively, one may think of

Fig. 1 Fibonacci spiral (*source* [12])

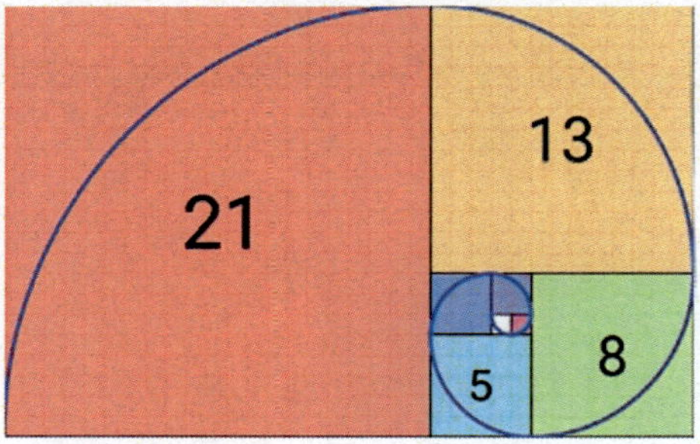

joining the opposite corners of these squares by a straight line. Then the length of this diagonal line segment would be the square root of the side of the square. The partial lengths with this arrangement are: 5.15, 18.40, 53.08, 143.88 and 381.58. The partial spiral length at the end of the i-th turn in the case of smooth curve is 6.85 times that at the end of its previous turn. It is 2.6 times in the case of the spiral obtained by joining the corners by straight lines. It's a matter of normalization, what we shall call unit and hence either of these could be mapped to the expected partial spiral lengths getting doubled at the end of each turn. So, we imagine the topology of the shruti cycle as a Fibonaaci spiral. The ratios 1/1, 2/1, 3/2 and 5/3 and 8/5 are the ratios of successive Fibonacci numbers for the first 6 numbers in the series.

Plenty of examples including the beautiful arrangement of compound leaves of many plants in our surroundings naturally follow the Fibonacci series. Fibonacci is known to be a generator of rich aesthetic patterns. Natyashastra has references to Fibonacci numbers in the context of the verse meters [8]. It is most likely that the shrutis are derived from the same logic. Given this, not only 3/2, all the 5 ratios are fundamental to HCM. 1/1 is a handle to continue with a particular note and 2/1 is for obtaining the same note in the adjacent upper octave. If you don't want to jump over a complete octave, rather if you wish to break at around half of the distance then 3/2 is available. 5/3 is a variant of dha which is a place to stop between pa and sa. 8/5, another variant of dha is yet another place when you are at dha and completing the octave.

<u>C4</u>: The 22 shrutis were generated using Fibonacci numbers.

We construct a device as follows Fig. 2:

When you are at Sa, using Fibonacci ratios you may explore four shrutis, 3/2, 5/3, 8/5 and 4/3 (Fig. 2a). Replace sa with ma and get the next set 2/1, 10/9, 16/15 and 16/9 (Fig. 2b). sa is redundantly generated (marked with a violet rectangle). Therefore, we get three new shrutis. Next, we replace ma with ni and so on. The roots of the trees and the shrutis each of them generate are listed in Table 5.

This is a simple mnemonic device. With respect to each root what we compute is its pa, two variants of dha and ma. The bold marked figures in Table 5 are for sa and pa missing the tonic sa and ma corresponding to this pa. They need not be generated. The underlined figures are redundantly generated. The beauty of this method is that

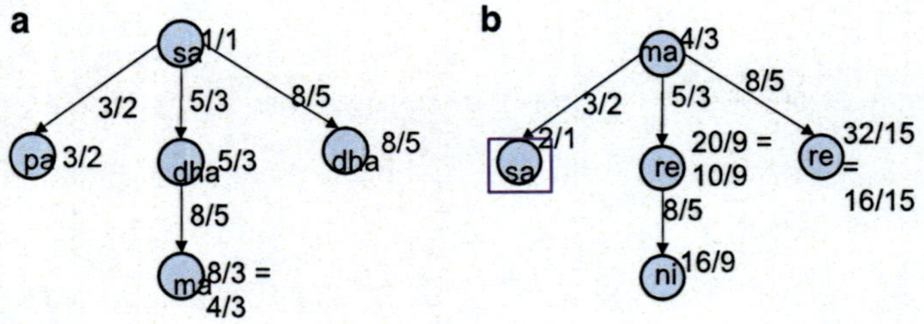

Fig 2 Fibonacci-based Shruti generator, 2 instances. **a** Root = sa. **b** Root = ma

Table 5 Computation of 22 shrutis using Fibonacci ratios

Tr#	Root	Shrutis	Tr#	Root	Shrutis
1	sa (1/1)	3/2, 5/3, 8/5, 4/3	6	pa (3/2)	9/8, 5/4, 6/5, 1/1
2	ma (4/3)	2/1, 10/9, 16/15, 16/9	7	Re (9/8)	27/16, 15/8, 9/5, 3/2
3	ni (16/9)	4/3, **80/54, 64/45**, 32/27	8	dha(27/16)	81/64, 45/32, 27/20, 9/8
4	ga (32/27)	16/9, **160/81, 256/135**, 128/81	9	ga(81/64)	243/128, **135/128, 81/80, 135/80**
5	dha(128/81)	32/27, **320/243, 512/405**, 256/243	10	Ni (243/128)	729/512, **405/256, 243/160**, 81/64

at any point it shows three possibilities to move on. Even the ratios that are not considered to be valid could be considered for the points for exploring further and one may get valid shrutis. For instance (80/54) * (3/2) = 10/9.

The path to explore the shruti set is the performer's choice. If by any chance this exploring characteristic resembles the musicians' intuition in the improvisation then this device could be evolved into an automaton for HCM. This is a pointer for further research and exploring the same is not within the scope of this work.

The machine is governed by the following rule: successive application of 3/2 is possible maximum 5 times while that of 5/3 and 8/5 is not allowed. It means that the ratio 3/2 is privileged. What could be the reason? Why are the shrutis 25/18 and 32/25 banned? Would it not lead to a symmetric structure? Is a spiral more aesthetically structured than a circle? Or is it because of its potential to grow infinitely like fractals? The latter sounds more reasonable. However, a few traditions seem to have acceptance to the shrutis generated by successive applications of 5/3 and 8/5 as well. For example, [7] mentions 45 shrutis. More research calls for analyzing their pros and cons. However, given that the Fibonacci sequences are known to the ancient Indian people, and Fibonacci being a source of aesthetic patterns in surroundings, it may be possible that the same set of 22 shrutis might have prevailed in HCM despite the changes in few other elements.

The shape of the Fibonacci spiral does exactly match with that of the Sanskrit alphabet ga in Devanagiri script, (the shape excluding the vertical bar and the top line in ग). We speculate that the Fibonacci-based scale was given the name "Swayambhu ga chi" for its shape and the ga in it has over the time been mistaken for the gandhar that it contains. Actually, we saw that it is dha that serves the basis to develop this scale. Also, it is confirmed that tuning of a musical instrument cannot be completed only with ga, it has to be at least for dha and ga though the process is called gandhar-tuning. Yet, when the tonic sa is performed, it gives rise to not only ga, it generates all other notes too. In the light of these oversighted facts, validation of our speculation could be proposed as a research problem. It may add a few newer perspectives to the foundations of HCM.

Conclusion

We attempted a logical intuitive visualization for computing the set of 22 shrutis. A perception regarding the 22 shrutis of the musicians of the early twentieth century has been developed. A Fibonacci sequence-based computation of 22 shrutis is claimed as an important contribution of this paper. We hope that the concept would be more researched and evolved into an automaton for modern HCM. Several small but important issues like the reason behind sa and pa being invariants, the ratio of 80/81 getting called as syntonic comma, the evolution of swaymabhu gandhar, the fundamental ratios 3/2, 5/3 and 8/5, and not 5/4, etc., have been addressed without making any undue assumptions. In short, the material presented in this paper should give a fresh perspective for the researchers to look into the mathematical foundations of HCM.

Acknowledgements Author takes this opportunity to remember the debt of the Late Dr Bharatee Vaishmapayan, the Late Mr Sadashiv Navangule and all the teachers in the Department of Music and Dramatics at Shivaji University, Kolhapur, for introducing her the world of HCM very patiently and passionately. Special thanks are due to Dr Jayant Kirtane, a keynote speaker at COMAD'19, for introducing the problem of computing 22 shrutis to the audience and generating inputs for our ideas. Thanks to Mr Srijan Deshpande for providing several clarifications, pointers and references that immensely helped in organizing the paper.

References

1. Bharatmuni, Natyashastra, https://sanskritdocuments.org/sanskrit/major_works/. Accessed 5 Sept 2022
2. V. Bhatkhande, Hindhustani Sangeet Paddhati, 2nd edn. (Popular Publication, 1995)
3. S. Deshpande, Sangeetache archives, Kalanirnaya diwali issue (2007)
4. J. Kirtane, Epistemology of intonation, Keynote speech at COMAD'19
5. C. Kunte, Preface to the volume of Maharashtra Nayak on music (Saptahik Vivek, 2019)
6. V. Oke, 22 shruti (Sanskar Publications, 2010)
7. R. Sane, Swara aale Kothooni, http://rajeevsane.blogspot.com/2015/07/blog-post_89.html. Accessed 5 Sept 2022
8. P. Singh, *The so-called Fibonacci Numbers in Ancient and Medieval India* (Historia Mathematica, Elsevier, 1985)
9. D. Thakur, *The Notion of Twenty Two Shrutis - Frequency Ratios in Hindustani Classical Music* (Resonance, Springer, 2015)
10. https://en.wikipedia.org/wiki/Bilaval. Accessed 5 Sept 2022
11. https://en.wikipedia.org/wiki/Fibonacci_number. Accessed 5 Sept 2022
12. https://en.wikipedia.org/wiki/Fibonacci_number#/media/File:Fibonacci_Spiral.svg. Accessed 5 Sept 2022
13. https://en.wikipedia.org/wiki/Golden_ratio. Accessed 5 Sept 2022
14. https://en.wikipedia.org/wiki/Golden_spiral. Accessed 5 Sept 2022
15. https://en.wikipedia.org/wiki/Logarithmic_spiral. Accessed 5 Sept 2022
16. https://en.wikipedia.org/wiki/Ptolemy%27s_intense_diatonic_scale. Accessed 5 Sept 2022
17. https://en.wikipedia.org/wiki/Shadja. Accessed 5 Sept 2022
18. https://en.wikipedia.org/wiki/Syntonic_comma. Accessed 5 Sept 2022

Signal Processing in Music Production: The Death of High Fidelity and the Art of Spoilage

David Courtney

Introduction

Commercial music production shows a complex relationship between aesthetics and the technological state of the art [6]. In its early history, there was a correlation between the aesthetics of what the consumers wanted, and what the electronics industry was trying to deliver. From the early days up until the mid-twentieth century, the public wanted the maximum fidelity that the industry was able to produce. However, from the 1960s onward there arose an antagonistic relationship between the electronics industry's push toward greater fidelity, and the tastes of the consumer. There were about two decades of confusion as the field was struggling for a sense of direction. However, by the 1980s it became clear that further increases in fidelity were not a high priority. Consumers had other preferences. These were convenience, a processed sound which mimicked the sound of older technologies, and new sounds that had no counterpart in the real-world. This led to a variety of techniques for processing an audio signal during the production process.

The aesthetics of signal processing will be a major focus of this work. However, there will also be some discussion of the technical aspects of signal processing.

We can divide signal processing into two approaches; analogue and digital. It may surprise many readers to find that analogue techniques are still used. Yet for many applications, analogue techniques are still preferred. Unfortunately, cost considerations tend to limit these approaches to the higher end studios. Therefore digital approaches are very common [8].

D. Courtney (✉)
Houston, TX, USA
e-mail: david@chandrakantha.com
URL: https://www.chandrakantha.com

A. Salgaonkar and M. Velankar (eds.), *Computer Assisted Music and Dramatics*, Advances in Intelligent Systems and Computing 1444,
https://doi.org/10.1007/978-981-99-0887-5_14

But it isn't just the analogue/digital divide that we will discuss, because digital signal processing may be further divided. We can divide this into two approaches: frequency-domain and time-domain processes. These will be touched upon in this paper.

Aesthetics of Spoilage

There is a very common yet generally unappreciated aesthetic of spoilage. It begins when people are forced to consume commodities that have been marred or blighted in some way. Presuming that such commodities still retain a level of utility or nutrition, there is a tendency for people to become attached to them. It is at that point that people begin to intervene and take control over the spoilage and elevate it to that of an industrial process.

Examples of this aesthetic are almost too numerous to mention. Some of the most common ones are spoiled grape juice (wine), spoiled milk (yogurt/cheese), (Fig. 1) rotten tea leaves (brown tea as opposed to green tea), or faded and torn jeans. One could no doubt come up with many more examples. As we progress through this paper we shall see that similar aesthetic forces are at work in the production of music.

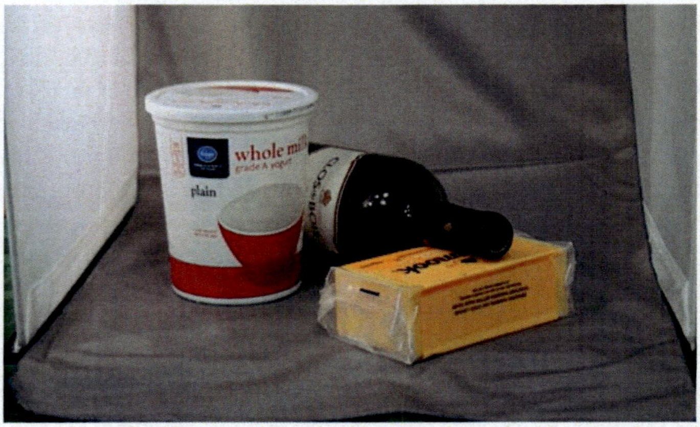

Fig. 1 "Spoiled" food

Aesthetics of Music Production

Aesthetics is inextricably linked to every aspect of music production. Although the academic treatment of the aesthetics of music is well represented [9] there has been very little attention given to aesthetics and contemporary music production. Fortunately, this is starting to be addressed, at least within Western musical genres [13]. Contemporary music production deals with three overall aesthetic approaches. There is the neutral sound, the classic sound, and the fantasia. Let us look into these three aesthetics in greater detail.

The Neutral Unprocessed Sound (Hi-Fi)

The aesthetic behind the neutral sound is the oldest; it actually represents an audio signal with no processing at all. It is based upon the principle that the finished recording should be as faithful to the original audio source as possible. This concept was known as "Hi-Fi" or "High Fidelity" in the 1960s and 1970s. However, in recent decades this term has been replaced by "neutral", "unaffected", or "uncoloured".

This aesthetic followed the trajectory of the development of recording technology. The early days of recording saw each new generation of equipment having better fidelity than the preceding one. Up until the middle of the twentieth century, this was the only aesthetic that guided the recording industry.

However, in the middle of the twentieth century, this approach lost favour with the public. The deviation from fidelity was driven by two things, the demand for convenience, and the desire for sonic experimentation.

As early as the 1970s, consumers made it clear that convenience was a more important consideration. The cassette, which was invented for the humble purpose of dicta-phones, became enormously popular. This was in spite of its poor fidelity. Fidelity began to reassert itself again in the 1980s and 1990s with the introduction of a plethora of high quality formats such as the DAT, CD, or Minidisk [10]. With the exception of the CD, these other formats enjoyed only limited acceptance. Even the high quality CD was ultimately displaced by the lower quality MP3.

Even though the market has rejected fidelity (the neutral sound) as an important aesthetic, it still has two uses. The first is in the area of acoustical research and testing where the data must be uncoloured by vagaries of the equipment. Another is in music production where many studios like to start with an uncoloured sound. From there, they proceed to process the sound in accordance to the desires of the producers and musicians.

The Classic Sound

The "classic sound" represents another common aesthetic, one that today requires a considerable amount of processing.

This classic sound is derived from multiple sources. Primarily these are the vacuum tube (thermionic valve), the audio transformer, and magnetic tape. In some cases (e.g., electric guitars) the speaker helped define the sound. All of these had non-linearities which plagued the recording industry, but went on to define the tastes of the public.[1]

To understand the origins of this aesthetic, we have to look back at the 1960s. The world was seeing a second and third generation of people who had little or no exposure to real musical instruments. Virtually their entire musical exposure had been filtered through the electronic media of radio, disk, and finally TV [12]. The weaknesses of the analogue technology of the day influenced their musical tastes. Consequently, efforts to improve fidelity after the 1960s were met with diminished enthusiasm. The public made itself very clear that the music "just didn't sound right".

The public's embrace of this classic sound caused studios to hold on to their old equipment. Manufacturers of recording and processing equipment continued to churn out "better" equipment. But they were feeling the pressure posed by the studios' reluctance to part with their older equipment. The commercial desire for classic equipment was at odds with the advancing technology.

One area that was particularly problematic was that of vacuum tubes (thermionic valves). The period between 1965 and 1975 saw a drastic reduction in their manufacture. By 1980 supplies of NOS (New-Old-Stock) tubes were dwindling and studios were feeling the pinch.

Relief from the shortage of vacuum tubes (valves) came from an unexpected source. The collapse of the Soviet Union in the 1980s created an opportunity for Western hardware manufacturers to find a permanent source of vacuum tubes. At least one musical equipment manufacturer went into the former Soviet Union, bought up tube manufacturing plants, modernise them, and used the tubes for their own equipment [4]. The most notable in this regard was Mike Mathews of New Sensor Corp. [7].

Another source of vacuum tubes came with the opening up of Chinese trade. The Chinese were very willing to step in and supply tubes to an eager Western market. Even today, Chinese manufacturers are major suppliers of vacuum tubes to the West (Fig. 2).

This paints a fairly clear idea as to the vacuum tube (valve) and its retention in high-end studio equipment; but what about the audio transformer.

It appears that Western manufacture of audio transformers did not suddenly drop, but showed a gradual shift toward China, Russia, and the former Soviet Republics. Today the transformer tends to be found only in high-end audio equipment. Although hardware is the preferred approach to obtaining the transformer sound, software emulation is also available [2].

The magnetic tape sound was much more difficult to maintain. In the past producers would record, edit, and process their work in the computer, record it to tape, then re-digitise the finished recording. This was very awkward. The tape sound was not practical until the introduction of software emulation. Today this is handled digitally with programs/processors from Empirical Labs, Avid, Electro-Harmonix, and a host of others.

[1] There were also the problems associated with the disk; the most significant being surface noise. However, this did not translate into a commonly sought after aesthetic.

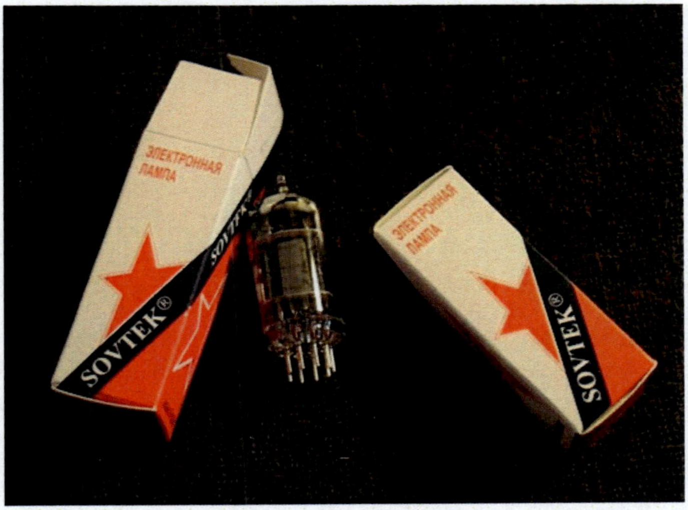

Fig. 2 Modern Russian made vacuum tubes

One characteristic of the classic guitar sound comes from the speaker. From a purely engineering standpoint, if one wants a good "clean" sound, one goes directly from the instrument into the mixing board. The idea is to keep the signal path as simple as possible. But this is rarely done. Normally one goes from the instrument to the amp/speaker. Then one simply places the microphone in front of the speaker [6].

The classic sound, like the neutral sound, is considered a base from which one works. However, where the neutral sound is not acceptable for a finished product, the classic sound is used in certain genres.

The Fantasia

It is a very common aesthetic for an engineer to create or modify sounds in ways which do not exist in nature. Even though this is very common, there is no widely accepted term for it. For the purpose of this paper, we will borrow the term "fantasia" from the discipline of film studies.

The fantasia is a sound which has the greatest degree of processing. This also goes back to the 1960s. At that time, such effects were always implemented through analogue means. Names such as Joe Meek, Robert Moog, and Geoff Emerick, are but a few which stand out in this regard.

This aesthetic is very popular, but today is almost always handled digitally. It is amply represented by manufacturers such as Eventide and a host of others [2].

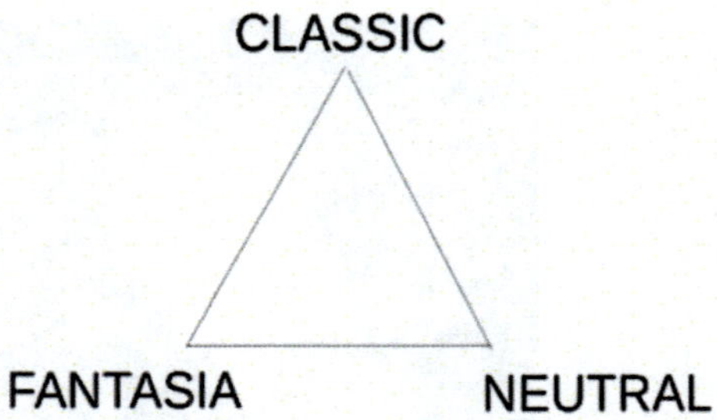

Fig. 3 The aesthetic map for which all recordings exist

Today's Aesthetic Map—These three aesthetics, the neutral (Hi-Fi), the classic, and the fantasia, form an aesthetic map which encompasses all musical production. It is one which has three poles as shown in Fig. 3.

Analogue Approaches to Signal Processing

Some level of analogue signal processing is unavoidable. There is always an analogue signal path that starts at the source of the sound and ends in the digital interface of the computer. The decisions that we make regarding this chain will have a profound effect upon the finished product.

It begins with one's choice of microphone. Does one go for a relatively neutral microphone, or a heavily affected one? Does one go for a ribbon microphone, a condenser microphone, or a dynamic one? [3]. If one chooses a condenser microphone, should it be solid state or vacuum tube?

After the microphone, the subsequent signal path must be considered. What kind of pre-amplifiers and other effects will the signal pass through before being sent to the interface? Does one pass the signal through an equaliser and/or a compressor? How does one's console (desk) behave?

But analogue processing is not limited to the initial recording phase of production; it may be used in the mix-down as well. High-end digital audio workstations allow an analogue processor to be patched into the virtual signal path. This is accomplished by sending the signal out of the computer via one's interface, into the analogue effects unit, and then back into the computer via the interface.

Digital Approaches to Signal Processing

Digital approaches to signal processing may not be the best for every purpose, but they are the most common. Where the vacuum tube and transformer may still be used as analogue processors, many other effects are only practical with software. This is especially the case with many modern effects that could not have even been imagined a few decades ago [2] (Fig. 4).

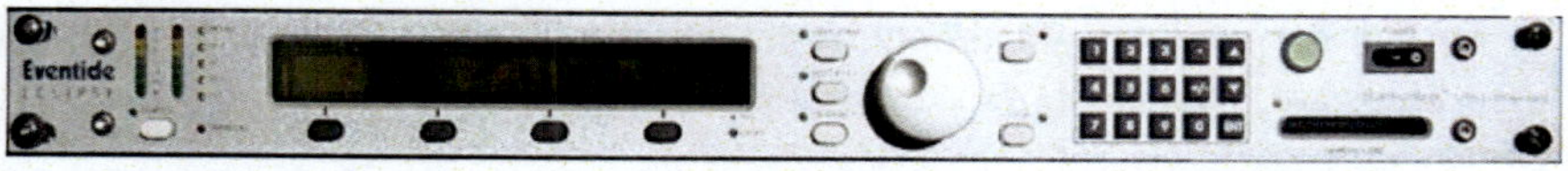

Fig. 4 Typical high-end digital effects processor (Eventide Eclipse)

Real world conditions do not require studio engineers to deal with the mechanics of digital algorithms. For the most part, these effects are mere "black boxes" into which the engineer feeds the signal and gets a desired sound out.

However, when one is programming such effects, the mechanics becomes very important. Audio effects programming is a demanding job, because psychoacoustic and aesthetic considerations are thrown into the mix. These will normally supersede any of the usual programming considerations.

But let us assume that the reader is going to program their own effect. There are the obvious considerations of programming language, hardware environment, and whether the effect is going to be implemented real-time or not. Space considerations do not allow us to even touch upon this topic. But let us assume that all of these have been considered. Let us further assume that we have a basic idea as to how the effect is supposed to sound. What is the next step?

One of the first things to consider is whether we process a digital effect in the time-domain or the frequency-domain?

Inverse Domains

Time and frequency domains represent two different ways of dealing with acoustical events. Just as a coin has two sides, so too do these two approaches offer equally valid, but different, ways of looking at the same event [5].

Time and frequency domains are not equal in terms of dealing with the various aspects of sound. Some applications are more easily handled in the time-domain, while others are more easily handled in the frequency-domain. Choosing which domain to perform a particular function is the first and perhaps most important decision when developing a digital audio effect.

Time-Domain Processing

Time-domain processing is a very common approach. For this, the digitised signal is placed in tables. The tables are then manipulated in whatever fashion we want. The modified tables then become our processed signal.

The time-domain approach is very popular for many applications. It is the most direct way of processing and is the least computationally intensive. It is most suitable for delay, echo, reverberation, flanging, and time shifting. One can even do pitch shifting in the time-domain simply by resampling the data. Because time-domain algorithms are so efficient, most applications can be executed in real-time.

Frequency-Domain Processing

Frequency-domain processing is computationally intensive; but due to the increasing power of modern digital equipment, it is rising in popularity. In this approach the signal is digitised, the data is then converted to frequency-domain by a Fourier transform. It is then written to tables. The tables are then manipulated according to one's desires, then converted back to time-domain by an inverse Fourier transform. This becomes our processed signal.

The Fourier transform is an important consideration in creating any usable frequency-domain based effect, but these come in different versions. The most commonly used version for audio effects processing is the Short Time Fourier Transform (STFT). This may be visualised as a small window which slides along the continuously varying musical signal. It is a very close relative of the Fast Fourier Transform (FFT), which in turn belongs to the family of the Discrete Fourier Transform (DFT).[2]

There are a number of processes which are well implemented within the frequency-domain. These are pitch shifting, harmonising, thickening, sparkle, and similar effects.

Both Domains Together

It is common to perform different aspects of an effect in different domains [11]. For instance, one may take a batch of samples and use time-domain techniques to identify a period (i.e., pitch detection). One then takes those samples and performs a Fourier transform. These are then manipulated with the appropriate frequency-domain algorithms. After that, one then performs an inverse Fourier transform to bring the data back into the time-domain. A few finishing time-domain touches usually completes the effect.

[2] There is an alternative to the Fourier transform which seems to have some level of use; this is the FWHT (Fast Walsh-Hadamard Transform). Unfortunately, due to the high level of secrecy surrounding commercial software, I do not have any information about this. However we can only surmise that this is more applicable to data compression rather than actual musical effects.

Conclusion

We have seen in this paper that aesthetic and commercial pressures have engendered some very bizarre pieces of hardware and software. This has had profound philosophical and practical implications due to its complete rejection of the trajectory of audio technology that was in place since the earliest days. This teleology was originally aimed at ever increasing fidelity. But by the middle to latter part of the twentieth century, the emphasis shifted to artistic manipulation of source material rather than faithful reproduction. The result is that audio software and hardware developed unique approaches, many of which are baffling to traditional electronic engineers.

References

1. Empirical Labs, *F.A.T.S.O Users Manual* (Empirical Labs, West Milford, NJ, USA) (undated)
2. Eventide Audio Inc., *Eventide Eclipse User Manual Manual/Software release 4.0.2 updated May 8, 2018* (Eventide Inc., Little Ferry, NJ 07643 USA, 2018)
3. D.M. Huber, R.E. Runstein, *Modern Recording Techniques*, 6th edn. (Focal Press, New York, 2005)
4. A.E. Kramer, *From Russia with Dread* (New York Times, 2006). https://www.nytimes.com/2006/05/16/business/worldbusiness/16cheat.html
5. A. Lerch, *An Introduction to Audio Content Analysis: Applications in Signal Processing and Music Informatics* (Wiley-IEEE Press, 2012)
6. H. Massey, *Behind the Glass: Top Record Producers Tell How They Craft the Hits* (Miller Freeman Books, Berkeley CA, 2000)
7. R. Murdocco, *A Brief History of Electro-Harmonix* (Reverb, Gear History, 2016). https://reverb.com/news/a-brief-history-of-electro-harmonix
8. J. Pakarinen, V. Välimäki, F. Fontana, Recent advances in real-time musical effects, synthesis, and virtual analog models. EURASIP J. Adv. Signal Process **2011**, 940784 (2011). https://doi.org/10.1155/2011/940784
9. R. Scruton, *The Aesthetics of Music* (Oxford University Press, New York, 1997)
10. A. Spanias, et al., *Audio Signal Processing and Coding* (A John Wiley and Sons, Hoboken New Jersey, 2005)
11. S. Tempelaars, et al., *Signal Processing, Speech and Music* (Routledge, London and New York, 1996)
12. A. Volmar, Experiencing high fidelity: sound reproduction and the politics of music listening in the twentieth century, in *The Oxford Handbook of Music Listening in the 19th and 20th Centuries* (Oxford University Press, 2019)
13. D. Williams, Tracking timbral changes in metal productions from 1990 to 2013. Metal Musical Stud. **1**(1), 39–68 (2014). Intellect, Bristol UK

Computational Indian Musicology: Challenges and New Horizons

Vinod Vidwans

Introduction

A significant number of computer professionals, musicologists, and practicing musicians are engaged in research in the domain of computational musicology, both domestically and globally. This domain is highly interdisciplinary in nature and demands expertise in both the fields of computational technology and music [15]. Being interdisciplinary in nature, computational musicology finds its place at the intersections of diverse fields viz. mathematics, sciences, social sciences and humanities. Music is also a cultural construct, deeply rooted in the history, tradition and ethnicity of society. Computational musicology, therefore, requires contribution of curiosities, questions, and insights from these fields. The canvas is quite large.

The broader objective of computational musicology is to study music using computational technologies. The study or investigation encompasses a wide range of dimensions of music that may include understanding the fundamental precepts of music from a computational perspective, structural analysis of music, the nature of music appreciation, the process of music representation as data and its retrieval [15], the process of music generation and its simulation [5]. This combination of computation, analysis, and statistics makes this field unique. In brief, computational musicology is a domain of knowledge that uses computational methods, models, and statistical analyses or informatics to develop insights into the phenomenon of music.

Computational musicology enriches musicians and musicologists to reflect upon their own ideas about music and understand them better from a more analytical perspective. It is observed that music practitioners and performers are averse to such analyses in general. Computational musicology provides a unique opportunity to

V. Vidwans (✉)
Professor and Former Chair, FLAME School of Fine and Performing Arts, FLAME University, Pune, India
e-mail: vidwans@flame.edu.in

A. Salgaonkar and M. Velankar (eds.), *Computer Assisted Music and Dramatics*, Advances in Intelligent Systems and Computing 1444,
https://doi.org/10.1007/978-981-99-0887-5_15

innovate and take their art to a different level. In turn, the domain of computation is enriched by the research into non-conventional problems.

It is necessary to take cognizance of the fact that computer researchers try to understand music from an analytical perspective. Such an analysis may result in a wide range of applications for the music industry, apart from developing deep theoretical insights. Musicians perceive music differently. Musician's concerns are centred around enhancing the quality of performance. Many of the musicians are oriented towards the pedagogy of music. Musicologists are more concerned about understanding the process of musical appreciation and the dynamics of composition. If computational musicology can significantly contribute in these regards, then such research is fruitful.

Current Status in India

Currently, research in Indian computational musicology is progressing in diverse directions. There are two main categories of such research. Firstly, most of the researchers are using contemporary methods, models and strategies in computer science to understand diverse aspects of music. Secondly, there are researchers who are using computational technology to investigate music within the traditional framework of Indian music. These two types are becoming increasingly prevalent which is evident from the nature and variety of other chapters from this book itself.

The first approach tends to develop novel computational perspectives without referring to the traditional framework of Indian music. This is evident from the research published in this book. This is elaborated in the ensuing paragraphs.

Information Retrieval: Musical information retrieval techniques are being used and innovated to capture significant features of a musical performance and then characterize fundamental musical concepts such as melody or musical modes by some of the authors in this book.

Machine Learning: Some other researchers are interested in feature extraction paradigm to understand the significant musical phrases or aspects of music using a machine learning approach.

Optimization Techniques: There are researchers who are engaged in using various optimization techniques to generate compositions.

FSA Models: For quite some time many researchers have been exploring the finite state automaton (FSA) paradigm to generate Indian music.

Computer Assisted Analysis: There is a trend which uses sophisticated software and computer tools for analysing pre-recorded performances of master performers to develop insights about the fundamental concepts and traditional terminologies used in Indian music.

Semiotic Analysis: Certain other researchers are interested in the semiotics of music. They are trying to extract the semantic information from the non-verbal components such as tempo, melody, instrumentation and orchestration.

It appears from these approaches that researchers have their own perspectives and motivations to conduct research in Indian music using computational models, methods and other software technologies. The research is being conducted primarily through individual efforts with little substantial institutional support. It is an encouraging sign that some serious efforts are being made in taking the field of computational musicology ahead against all odds.

The second approach adheres to the traditional framework. The historical context and future directions of research which use this approach are elaborated in the following paragraphs.

Indian music has a long history and tradition. The earlier musical practices are encoded in three important treatises called *Naradiya Shiksha*, (Narada, date not known), *Natyashastra* (Bharata, date not known), and *Sangit Ratnakar* (Sharangdeva, 13th c.). India also has a well-developed system of codification of knowledge known as the '*Sutra*' system. These treatises are considered as '*Sutra*' treatises for Indian music. Sutra treatises are codified texts on specialized subjects or topics. The codifications are highly precise and occasionally cryptic. Indian musicians and musicologists have great respect for these treatises. Indian computational musicology needs to position itself in the framework of these treatises. Some efforts have made progress along these lines. Though this research is not very strictly following the traditional framework, it is omnipresent in these efforts.

Raga Recognition: *Raga* recognition is an area of curiosity for researchers. Well-informed attempts are being made to utilize the available technologies and software systems to extract salient features of a construct called *Raga* and develop theoretical schema based on similarities and differences in the musical and tonal material.

Computational Choreography Generation: A noteworthy effort in automatic generation of innovative choreography based on the concept of '*Nritya*' from Bharata's *Natyashastra* needs special mention. In such a work we find a hybrid approach which uses vector representations of bodily movements, genetic algorithms and use of fractal aesthetics to achieve the universal metrics for aesthetic computation. Such efforts focus on evaluating whether aspects of art such as dance aesthetics are amenable to computation.

Computational Music Generation: The author of this chapter has developed '*AIRaga*' and '*AITala*' systems based on Bharata's *Natyashastra*. Both the systems use hybrid strategies such as optimality theory (OT), generative grammars, and genetic algorithms. Both the artificially intelligent systems generate innovative output without human intervention. The *AIRaga* system generates a *Bandish* or a musical composition in the given *Raga* and renders it with artificial instruments into a playable composition. The *AITala* system generates a composition in a given *Tala* and renders it. These systems do not rely on any database or corpus of existing *Ragas* or *Talas*. Being generative systems, both the systems generate the data required for music generation based on the encoded expert knowledge they possess (Vidwans, V. 2008). The detailed description of these systems is beyond the scope of this chapter.

Computational Indian Musicology

It is evident from the above discussion that computational musicology borrows methods, models and statistical analysis strategies from the field of computer science and allied disciplines while the content and themes of research are taken from the domain of music. Since Indian music has a long tradition and there exists substantial codified knowledge it can be argued that research in Indian computational musicology ought to be anchored in traditional frameworks. This will provide strong foundations to computational research. It is necessary to take into account the fact that Indian music is quite different from Western music in its theory, structure, and expression. Computational models, strategies and approaches used for Western music may not work for Indian music. If one gets deep into the Indian music, theoretical constructs and their practice/performance, one realizes that Indian music needs indigenous models and strategies based on the Indian theoretical paradigms. Indian music cannot be studied in isolation. It can be defined as a conglomeration of dance, vocal music, and instrumental music including percussion. Poetry and lyrical structures also impact music to a greater extent. Bharata's *Natyashastra*, therefore, looks at music as a part of a larger domain of theatrical arts that includes all the above aspects. It entails that for Indian computational musicology, one needs to adopt an inclusive approach by referring to treatises on Indian theatre (*Natyashastra*), Indian poetics (*Kavyashastra*), Indian prosody (*Chhandashastra*), and Indian aesthetics or the theory of *Rasa*.

Already categories for musical research are well-established [15]. The categories such as codified music theory and practice, historical musicology, ethnomusicology, and aesthetics of music are already extant and are serving their purpose to a significant extent as far as conventional research is concerned. While preserving the themes and content of these categories, new nomenclatures ought to be innovated to augment the computational perspectives and connect them to traditional Indian frameworks. This is probably the right time to contemplate the development of a taxonomy of computational musicology in an Indian context. Thus, a taxonomy for computational Indian musicology can be envisaged and would broadly cover the following areas: Indian music theory and computational analysis, computational performance research, computational historical (musical corpus study), musicology, computational ethnomusicology, cognitive musicology and computational aesthetics, and finally computational music generation. In the following section, it is discussed how these categories within musicology may contribute to computational musicology and how various issues in these areas may be addressed.

Computational Music Theory and Analysis

Computational analysis of musical pieces has been a central theme in computational musicology the world over. Indian music is known for its codified analytical structure as described in treatises. Basic musical concepts such as *swara*, *shruti*, *saptaka*, *grama* and others need to be redefined in order to be applicable to computational music analysis. The determination and characterization of *shrutis*, *swaras*, *swara*-modulation, *gamaka*, *aalap*, *taana* and musical phrases and segmentation of musical pieces need to be analysed from a computational perspective. Tonal relations such as *Shadja-Panchama Bhava* and *Shadja-Madhyama Bhava* belong to the most widely studied musical concepts in Indian classical music. However, these technical concepts of *swara* relations as described in these treatises and in theory are often not associable with the concrete tonal relations occurring in musical performances in a scientific and precise manner. To achieve this correlation, computational models of tonal relations, based on theory, need to be applied to the digitized musical pieces to show the link between theoretic concepts and their application in actual performance. In the case of *Raga* recognition and generation, there is a possibility of using Artificial intelligence (AI) techniques or hybrid computational paradigms to enrich theoretical understanding. Therefore, formalization of these concepts is the first task in this direction. The second task is to digitize musical pieces in a professional way either by using sophisticated tools or by developing indigenous tools to achieve the same.

Another important area of computational approaches to Indian music analysis could be the study of *Tala* (Rhythm) and *Chhanda* (metre). *Tala* and *Chhanda* can be considered as one of the most prominent research areas for Indian Music. The automatic detection of the *Tala* and *Chhanda* structures through machine learning techniques may be fruitful. Rule-based approaches, genetic algorithm, Bayesian approaches or allied pattern detection algorithms may prove promising in this regard. Computational musicology certainly helps in linking abstract music theories with concrete performance analyses of musical pieces. It helps to verify implicit musicological knowledge and open up new vistas in the domain of music analysis.

Computational Historical Musicology

The goal of computational historical musicology in Indian context would be to understand the evolution of musical concepts from the *Vedic* era to contemporary music. Traditionally, the focus in historical musicology has been to collect original resources on musical treatises and manuscripts and study them closely to understand the patterns of development of music in India. Understanding the evolution of the *Ragas* and musical modes has been the primary focus. Apart from that, studying the biographies of great masters (especially the *Gharana* system in North India) and making a repository of *Raga/Bandish* (musical compositions) from various sources has been a major activity in recent times. Historical musicology always had the underlying motivation to show how the phenomenon of music as an autonomous, timeless aesthetic creation emerged. Music institutes and Universities have built some good repositories in this regard and are conducting research. Apart from those motivated individuals have their own collections. There exists a huge collection of scholarly publications and articles on Indian music. However, historians need to be trained and develop acquaintance with computational technology. There should be collaborative efforts between computer researchers and historians.

Computational Ethnomusicology

India has a very rich and diverse tradition of folk and tribal culture. A central theme in Indian computational ethnomusicology, in which statistical methods and approaches may be applied, could be the study of folk songs inherited by performers as an oral tradition. Oral traditions have subtle nuances of musical significance. In every generation, due to the process of oral transmission, new variations are introduced in the folk songs. For ethnomusicology, this is a very important phenomenon. Computational strategies are convivial with regards to documentation, preservation, and analysis of such variations in a systematic manner. Another important issue in ethnomusicology is the classification of traditional folk songs or tunes. It is believed that Indian *Ragas* have evolved from folk tunes but there are very few well-informed studies in this regard. This is a very important issue from an ethnomusicological perspective. The structure of *Raga* is highly sophisticated whilst the folk tunes are natural unbridled expressions of the tribals. Computational ethnomusicology can shed light on such issues. Computational approaches allow the processing of large data and information about folk tunes and folk songs of highly complex nature that may include historical information along with aural and musical information. Indian ethnomusicology is a data-rich field. Computational modelling of this data can help address some outstanding qualitative questions and issues in the field.

Computational Cognitive Musicology

Cognitive musicology studies the mental process behind the musical perception, how we understand music and the mental processes that lead to musical expression. While in the context of Indian music, a musicologist's concern is to investigate complex relations between *swaras*, *swara*-phrases and *Raga* performance, the cognitive musicologist is interested in how *swaras* and *swara*-phrases are perceived, experienced and processed by mind as a part of the musical appreciation and the listening process in general. Cognitive psychology uses computational tools to understand the human mind. In the context of Indian music, computational models may be useful, for instance, to understand the process of *Tala* interpretation while listening to percussion performance or to understand the mental processes behind the recognition of *Raga* or to understand the chunking processes responsible for segmentation of *swara*-phrases to compare *Ragas* or relate *Ragas* with each other. Empirical studies can be conducted to build computational models for testing some of the basic and foundational assumptions of traditional musical theories. Empirical as well as simulation studies can also be conducted to understand the perception processes behind the perception of *swaras,* perception of *Ragas*, recognition of *Ragas* and distinguishing Ragas from one another. Computational models can be useful to understand the nature of musical genius and judge how much is the role of intuition, reason, logic in music creation, presentation and appreciation. This can be further extended to computational aesthetics of music and computational semiotics of Indian music.

Computational Performance Research

Computational models can be built to understand the stylistic variations of performers. The same *Bandish* (musical composition) is presented differently by different performers. The same *Raga* is rendered differently at different times by the same performer or by different performers with variations. Computational informatics can throw light on the degree of variance in all such instances. This may lead to a well-informed understanding of the '*Gharana*' system in North India or a deeper level of understanding of individual styles of the performers. Stylistic analysis of musician's performance appears to be a promising direction for Indian music. In this regard, new software tools can be developed to achieve this goal. There is a tremendous scope for computational analysis as well as software development for Indian musical presentation studies and representation of performance through digital documentation and notation.

Challenges and the Way Forward

Computational musicology is opening up new avenues of analytical and empirical studies for Indian musical concepts through computational simulations in many different aspects of music. This mainly helps to unravel the secrets of implicit Indian musicological knowledge possessed by traditional musicians and performers. Computational investigations help us formalize foundational concepts of Indian music like *swara* and constructs like *Raga* to establish new research areas and take the domain of Indian music forward. Computational models help connecting different branches of Indian musicology across regions and connect past withthe present. Efforts in formalization and empirical testing of musicological concepts through model building and/ or simulation are very much relevant for the growth of Indian music.

Indian tradition considers music as an art however, it is also understood as a *shastra* or a discipline. A major challenge for computational musicologists is to translate artistic notions into scientific terminology. To apply computational methods, models or algorithmic processes to music, the prerequisite is that the subject matter of music needs to be broken down into the smallest possible units. Such a data will come in the form of audio signals or in the form of numeric data, and to arrive at this granular stage, computational musicologists will have to struggle a lot because as it stands the Indian musicians rarely nurture such an outlook about music. The computational musicologist needs to be musically inclined or have a profound understanding of the form of Indian music that one wants to investigate. The whole process of formalization of Indian music is fraught with uncertainties and frustrating cul-de-sacs.

The musicologist's knowledge will determine which aspects of Indian music can be deconstructed by existing formal models and the necessity of new formal models. As it is known, traditional Indian music, in principle, is deconstructed up to the level of 22 *shrutis* (microtones). But there are many theories which need to be verified and validated. These have been hinted at in academic papers (such as [6, 4, 8, 10, 14]). *Shadja-Pancham Bhava* and other principles are well-established in Indian theory, however, to build computational models around them needs a rigorous treatment accompanied by demonstrable implementations of the same. It is well understood and accepted that musical compositions are complex entities where simple linear approaches will fail miserably. The integration of diverse aspects such as the fundamental elements of music and the variety of rules that govern composition involving (but not restricted to) poetics, aesthetics and literary domains makes the task complex.

Composers and the audience have a high-level shared vocabulary of music. North Indian musicians and their audiences understand *Ragas* in a specific way whilst Carnatic musicians and their audiences may perceive *Ragas* in a different manner. Translating such a vocabulary for formalizations of music poses issues related to sociocultural contextualization, subjectivity of musicians and their audiences, and the emergence of certain musical genres due to socio cultural and geopolitical specificities. There are no simple solutions to these issues. Along with the issues related to musical consonances and dissonances, *Raga* and *Tala* complexities, presentational styles with *swara*-ornamentations, and *gamakas* multiply the complexity of computational analysis and model building. These are non-trivial issues because all these components ultimately contribute towards the genesis of musical semantics/semiotics of Indian music. The current research efforts hold a lot of promise in this regard.

References

1. Bharata (Date not known). Natyashastra of bharatamuni, with the commentary Abhinavabharati by Abhinava Guptacharya (Chapters 28–37), Vol. IV, edited by Kavi, M. Ramakrishna and Pade, J. S., Baroda: Oriental Institute Baroda, (1964)
2. Bhise, Usha R. (Ed.), Naradiya Shiksha, Original commentary by bhatta shobhakara, critically edited with translation and explanatory notes in English by Dr. Bhise, Usha R. Bhandarakar Oriental Research Institute, Pune, India (1986)
3. A. Brihaspati, Natyashastra, 28th Chapter—Swaradhyaya, (Hindi) with Sanskrit and hindi commentary (Brihaspati Publication, New Delhi, 1986)
4. A.K. Datta, et al, Objective analysis of Shrutis from the vocal performances of hindustani music using clustering algorithm, in the journal 'Ninad', Vol. 20, Kolkata: ITC Sangeet Research Academy (2006)
5. E. Coutinho, M. Gimenes, Jo˜ao M. Martins and Eduardo R. Miranda., *Computational musicology: An artificial life approach* (IEEE Publication, Portuguese conference on artificial intelligence, 2005)
6. Erlich, Paul, tuning, tonality, and twenty-two-tone temperament, originally published in xenharmonikôn 17, An Informal Journal of Experimental Music, San Francisco: The Just Intonation Network, San Francisco, USA; revised in 2002 (1998)
7. Kaundinyayan, Aamodavardhana et al, Naradiya shiksha, critical commentary, translation and explanation in sanskrit and hindi by shivaraj acharya kaundinyayan, varanasi: Chaukhamba vidyabhavana (2013)
8. M.M. Komaragiri, On the applicability of the ancient śruti scheme to the current fixed-tonic, variable-interval mēla system, in the journal of the madras music academy, vol. 81, (Chennai, India, 2010), pp.135–143
9. Narada (Date not known). Naradiya shiksha, original commentary by bhatta shobhakara, critical commentary and explanation in sanskrit by narayanaswami dikshit, edited by Satyavrat Sharma, Mysore: Rajakiya Shakha Mudranalaya, published in 1949
10. Rao, Suvarna Lata, Meer Wim van der, The Construction, Reconstruction, and Deconstruction of Shruti', Hindustani Music–Thirteenth to Twentieth Centuries, by Bor Joep et al (eds.), pp. 673–696, New Delhi: Manohar Publishers & Distributors (2010)
11. Sharangadeva, Nishshanka (13thc. AD). Sangita Ratnakar (original Sanskrit text), translated by Taralekar, G. H. (Marathi Trans.), Mumbai: Maharashtra Rajya Sahitya Sanskriti Mandala, Published in 1975
12. Thakar, Sulabha, Bharatiya Sangita Shastra: Nava Anvayartha (Marathi), Solapur: Suvidha Prakashan (2002).
13. V. Vidwans, (2008) http://computationalmusic.com/
14. V. Vidwans, The Doctrine of *Shruti* in Indian Music, FLAME University Publication. (2016) http://computationalmusic.com/the-doctrine-of-shruti-in-indian-music.php
15. F. Volk Anja, Wiering, P. Kranenberg, Unfolding the potential of computational musicology, Journal, R.J, Liu, K., & N.R. Faber, (eds.), in *Proceedings of the Thirteenth International Conference on Informatics and Semiotics in Organisations—Problems and Possibilities of Computational Humanities*. Fryske Akademy, Netherlands (2011)

Author Index

R. Lee (ed.), *Computer and Information Science 2021—Summer*,
Studies in Computational Intelligence 985,
https://doi.org/10.1007/978-3-030-79474-3